# Extrusion Bioprinting of Scaffolds for Tissue Engineering

Daniel X. B. Chen

# Extrusion Bioprinting of Scaffolds for Tissue Engineering

Second Edition

 Springer

Daniel X. B. Chen (iD)
University of Saskatchewan
Saskatoon, SK, Canada

ISBN 978-3-031-72473-2          ISBN 978-3-031-72471-8    (eBook)
https://doi.org/10.1007/978-3-031-72471-8



This Springer imprint is published by the registered company Springer Nature Switzerland AG
The registered company address is: Gewerbestrasse 11, 6330 Cham, Switzerland

If disposing of this product, please recycle the paper.

*To Qi Huang and Angel Chen*
*To Peter and Arlene Block*

# Preface

Advances in both engineering techniques and the life sciences have evolved extrusion bioprinting from a simple technique to one able to create diverse, yet complicated, tissue scaffolds or constructs for various tissue engineering and modeling applications. This has also led to several edited books that review and report these advances and developments. Emphasizing the advances of various bioprinting technologies, these books are generally written for experienced researchers in this field. In addition, no books have been dedicated to extrusion bioprinting technologies yet they are the most common way to fabricate scaffolds among the various bioprinting technologies available. Aiming to fill these gaps, this book introduces readers to the theory and practice of extrusion bioprinting of scaffolds for tissue engineering and modeling. The text emphasizes the fundamentals and practical applications of extrusion bioprinting for scaffold fabrication, in a manner particularly suitable for those who wish to master the subject matter and apply it to tissue engineering and modeling. Readers will learn how to design, fabricate, and characterize tissue scaffolds created by extrusion bioprinting technologies.

After an overview of tissue engineering, Chap. 1 provides a brief introduction to the development of scaffolds for tissue engineering and modeling as well as various scaffold fabrication techniques. Chapter 2 presents the general requirements imposed on scaffolds and the scaffold design process. Chapter 3 discusses the properties of biomaterials important for extrusion bioprinting and the hydrogels commonly used. Chapter 4 focuses on the common methods/techniques used to measure and characterize the mechanical properties of native tissues and scaffolds. Chapter 5 presents the fundamentals of preparing biomaterial solutions or bioinks with living cells for bioprinting scaffolds, while Chap. 6 focuses on the basic principles of extrusion bioprinting, along with the influence of bioprinting process parameters on scaffold fabrication. Chapter 7 discusses bioprinting-based and other approaches to create vascular networks within tissue scaffolds to facilitate their functions. The last chapter (i.e., Chap. 8) introduces the concept of controlled release and the common strategies for regulating biomolecules in tissue engineering and printed scaffolds.

## New to This Edition

The following list describes the key changes in the second edition.

1. A new chapter titled "Controlled Release of Biomolecules in Printed Scaffolds" has been added to reflect the recent advances in the field of bioprinting.
2. Important concepts (including printability) for extrusion bioprinting have been added or clarified based on recent literature.
3. All of the chapters have been revised with more examples, case studies, and illustrations to reflect the recent advances and developments.
4. End-of-chapter problems have been revised.


Saskatoon, SK, Canada                                        Daniel X. B. Chen

# Acknowledgements

I acknowledge the contribution of former and current graduate students and researchers in the Biofabrication Lab at the University of Saskatchewan, Canada, whose assistance and perseverance made the completion of this text possible. Specifically, in the first edition, Dr. Saman Naghieh co-authored Chap. 2, Dr. Fu You Chap. 3, Dr. Nahshon Bawolin and Dr. Nitin Sharma Chap. 4, Dr. Liqun Ning Chaps. 5 and 6, and Dr. Ariful Sarker Chap. 7. In the second edition, Dr. Amanda Zimmerling co-authored Chap. 8. My sincere appreciation also goes to Dr. Xiaoman Duan, Abbas Fazel Anvari Yazdi, Dr. Reza Gharraei Khosroshahi, Dr. Nitin Sharma, Dr. Lihong He, Kathryn Avery, and Xavier Tabil for their help revising the chapters in this second edition.

Daniel X. B. Chen

# Contents


**1  Extrusion Bioprinting of Scaffolds: An Introduction** . . . . . . . . . . . . . 1
   1.1  Introduction . . . . . . . . . . . . . . . . . . . . . . . . . . . . . 1
   1.2  Scaffolds for Tissue Engineering and Modeling . . . . . . . . . . . . 1
   1.3  Scaffold Fabrication . . . . . . . . . . . . . . . . . . . . . . . . . 5
      1.3.1  Traditional Techniques . . . . . . . . . . . . . . . . . . . . . 5
      1.3.2  Electrospinning . . . . . . . . . . . . . . . . . . . . . . . . . 6
      1.3.3  Three-Dimensional Printing . . . . . . . . . . . . . . . . . . . 7
   1.4  Extrusion Bioprinting of Scaffolds . . . . . . . . . . . . . . . . . . 9
   1.5  Advances and Limitations of Extrusion Bioprinting . . . . . . . . . . 11
   1.6  Summary . . . . . . . . . . . . . . . . . . . . . . . . . . . . . . . 12
   References . . . . . . . . . . . . . . . . . . . . . . . . . . . . . . . . 14

**2  Tissue Scaffold Design** . . . . . . . . . . . . . . . . . . . . . . . . . . . 17
   2.1  Introduction . . . . . . . . . . . . . . . . . . . . . . . . . . . . . 17
   2.2  General Requirements of Tissue Scaffolds . . . . . . . . . . . . . . . 17
      2.2.1  Architectural Properties . . . . . . . . . . . . . . . . . . . . . 17
      2.2.2  Mechanical Properties . . . . . . . . . . . . . . . . . . . . . . 21
      2.2.3  Biological Properties . . . . . . . . . . . . . . . . . . . . . . 23
   2.3  Scaffold Design Process . . . . . . . . . . . . . . . . . . . . . . . . 23
      2.3.1  Understanding the Composition and Organization
            of Tissue/Organs . . . . . . . . . . . . . . . . . . . . . . . . 24
      2.3.2  Designing Scaffolds with Appropriate Architectures . . . . . . 25
      2.3.3  Selection of Biomaterials/Cells . . . . . . . . . . . . . . . . . 26
   2.4  Typical Scaffold Structures by Bioprinting . . . . . . . . . . . . . . 29
   2.5  Summary . . . . . . . . . . . . . . . . . . . . . . . . . . . . . . . 32
   References . . . . . . . . . . . . . . . . . . . . . . . . . . . . . . . . 34

**3   Biomaterials and Bioinks for Bioprinting** . . . . . . . . . . . . . . . . . . . .   37
3.1   Introduction . . . . . . . . . . . . . . . . . . . . . . . . . . . . . . . . . . . .   37
3.2   Important Properties of Biomaterials for Bioprinting . . . . . . . . . .   38
    3.2.1   Printability . . . . . . . . . . . . . . . . . . . . . . . . . . . . . . . .   38
    3.2.2   Crosslinking Mechanisms . . . . . . . . . . . . . . . . . . . . . . . .   40
    3.2.3   Biological Properties and Biodegradation . . . . . . . . . . . . .   42
    3.2.4   Mechanical Properties . . . . . . . . . . . . . . . . . . . . . . . . .   44
3.3   Biomaterials for Bioprinting . . . . . . . . . . . . . . . . . . . . . . . . . .   45
    3.3.1   Natural Hydrogels . . . . . . . . . . . . . . . . . . . . . . . . . . . .   45
    3.3.2   Synthetic Hydrogels . . . . . . . . . . . . . . . . . . . . . . . . . .   50
    3.3.3   Composite Hydrogels . . . . . . . . . . . . . . . . . . . . . . . . . .   51
3.4   Summary . . . . . . . . . . . . . . . . . . . . . . . . . . . . . . . . . . . . . .   52
References . . . . . . . . . . . . . . . . . . . . . . . . . . . . . . . . . . . . . . . . .   53

**4   Mechanical Properties of Native Tissues and Scaffolds** . . . . . . . . . .   57
4.1   Introduction . . . . . . . . . . . . . . . . . . . . . . . . . . . . . . . . . . . .   57
4.2   Mechanical Testing Methods . . . . . . . . . . . . . . . . . . . . . . . . . .   57
    4.2.1   Basics of Mechanical Testing . . . . . . . . . . . . . . . . . . . . .   57
    4.2.2   Tensile and Compressive Tests . . . . . . . . . . . . . . . . . . . .   62
    4.2.3   Bending Tests . . . . . . . . . . . . . . . . . . . . . . . . . . . . . .   67
    4.2.4   Torsion Tests . . . . . . . . . . . . . . . . . . . . . . . . . . . . . . .   69
4.3   Viscoelastic Properties and Dynamic Testing . . . . . . . . . . . . . . .   72
    4.3.1   Viscoelastic Properties . . . . . . . . . . . . . . . . . . . . . . . . .   72
    4.3.2   Cyclic Load Tests . . . . . . . . . . . . . . . . . . . . . . . . . . . .   73
    4.3.3   Creep and Stress-Relaxation Tests . . . . . . . . . . . . . . . . .   73
    4.3.4   Frequency-Dependent Tests . . . . . . . . . . . . . . . . . . . . . .   75
    4.3.5   Mathematical Models of Linear Viscoelastic Behavior . . . .   76
4.4   Mechanical-Property Measurements of Native Tissues
    and Scaffolds . . . . . . . . . . . . . . . . . . . . . . . . . . . . . . . . . . . .   82
    4.4.1   Sample Preparation . . . . . . . . . . . . . . . . . . . . . . . . . . .   82
    4.4.2   Considerations During the Testing . . . . . . . . . . . . . . . . . .   83
    4.4.3   Case Studies: Measurement of Mechanical Properties . . . . .   84
4.5   Mechanical Properties of Scaffolds . . . . . . . . . . . . . . . . . . . . . .   94
    4.5.1   Influence of Scaffold Structure . . . . . . . . . . . . . . . . . . . .   94
    4.5.2   Influence of Scaffold Materials . . . . . . . . . . . . . . . . . . . .   95
    4.5.3   Time-Dependent Mechanical Properties . . . . . . . . . . . . . .   97
4.6   Methods to Improve the Mechanical Properties of Scaffolds . . . . .   98
    4.6.1   Use of Composite Materials . . . . . . . . . . . . . . . . . . . . . .   98
    4.6.2   Addition of Fillers . . . . . . . . . . . . . . . . . . . . . . . . . . . .   101
    4.6.3   Hybrid Structures . . . . . . . . . . . . . . . . . . . . . . . . . . . . .   102
4.7   Summary . . . . . . . . . . . . . . . . . . . . . . . . . . . . . . . . . . . . . .   103
References . . . . . . . . . . . . . . . . . . . . . . . . . . . . . . . . . . . . . . . . .   106

**5   Preparation of Biomaterial Solutions and Characterization
     of Their Flow Behavior** . . . . . . . . . . . . . . . . . . . . . . . . . . . . . . . . .   109
　5.1   Introduction . . . . . . . . . . . . . . . . . . . . . . . . . . . . . . . . . . . . . .   109
　5.2   Preparation of Biomaterial Solutions . . . . . . . . . . . . . . . . . . . . . .   109
　　5.2.1   Basics of Solution Preparation . . . . . . . . . . . . . . . . . . . .   109
　　5.2.2   Solutions with Living Cells . . . . . . . . . . . . . . . . . . . . . .   114
　　5.2.3   Solutions without Living Cells . . . . . . . . . . . . . . . . . . . .   114
　5.3   Flow Behavior Characterization of Biomaterial Solutions . . . . . . .   115
　　5.3.1   Flow Behavior and Its Classification . . . . . . . . . . . . . . . .   115
　　5.3.2   Flow Behavior Models . . . . . . . . . . . . . . . . . . . . . . . . . .   119
　5.4   Techniques to Characterize Flow Behavior . . . . . . . . . . . . . . . . .   124
　　5.4.1   Capillary Rheometer . . . . . . . . . . . . . . . . . . . . . . . . . . .   124
　　5.4.2   Cone-and-Plate Rheometer . . . . . . . . . . . . . . . . . . . . . .   127
　　5.4.3   Parallel Plate Rheometer . . . . . . . . . . . . . . . . . . . . . . . .   128
　　5.4.4   Oscillatory Rheometer . . . . . . . . . . . . . . . . . . . . . . . . . .   129
　5.5   Key Factors Related to Flow Behavior . . . . . . . . . . . . . . . . . . . .   132
　　5.5.1   Influence of Material Concentration . . . . . . . . . . . . . . . . .   132
　　5.5.2   Influence of Temperature . . . . . . . . . . . . . . . . . . . . . . . .   133
　　5.5.3   Influence of Cell Density . . . . . . . . . . . . . . . . . . . . . . . .   135
　5.6   Summary . . . . . . . . . . . . . . . . . . . . . . . . . . . . . . . . . . . . . . . .   136
　References . . . . . . . . . . . . . . . . . . . . . . . . . . . . . . . . . . . . . . . . . . .   138

**6   Bioprinting of Tissue Scaffolds** . . . . . . . . . . . . . . . . . . . . . . . . . . .   141
　6.1   Introduction . . . . . . . . . . . . . . . . . . . . . . . . . . . . . . . . . . . . . .   141
　6.2   Basics of Bioprinting Systems . . . . . . . . . . . . . . . . . . . . . . . . . .   141
　6.3   Flow Rates in Bioprinting . . . . . . . . . . . . . . . . . . . . . . . . . . . . .   144
　　6.3.1   Pneumatic-Driven Bioprinting . . . . . . . . . . . . . . . . . . . .   144
　　6.3.2   Screw-Driven Bioprinting . . . . . . . . . . . . . . . . . . . . . . .   151
　　6.3.3   Piston-Driven Bioprinting . . . . . . . . . . . . . . . . . . . . . . .   153
　　6.3.4   Effect of Needle Geometry . . . . . . . . . . . . . . . . . . . . . . .   154
　　6.3.5   Extrudability in Bioprinting . . . . . . . . . . . . . . . . . . . . . .   157
　6.4   Bioprinting of Scaffolds . . . . . . . . . . . . . . . . . . . . . . . . . . . . . .   158
　　6.4.1   Needle Movement in the Horizonal Plane . . . . . . . . . . . .   158
　　6.4.2   Needle Movement in the Vertical Direction . . . . . . . . . . .   161
　　6.4.3   Crosslinking in Bioprinting . . . . . . . . . . . . . . . . . . . . . .   162
　　6.4.4   Design of Bioprinting Parameters . . . . . . . . . . . . . . . . . .   164
　6.5   Profile and Structure of Printed Scaffolds . . . . . . . . . . . . . . . . . .   166
　　6.5.1   Profile and Structure . . . . . . . . . . . . . . . . . . . . . . . . . . .   166
　　6.5.2   Techniques to Characterize Profile and Structure . . . . . . . .   166
　6.6   Cell Damage in Bioprinting . . . . . . . . . . . . . . . . . . . . . . . . . . . .   170
　　6.6.1   Process-Induced Mechanical Forces . . . . . . . . . . . . . . . .   170
　　6.6.2   Cell Damage Due to Mechanical Forces . . . . . . . . . . . . .   174
　　6.6.3   Cell Damage and Its Characterization in Bioprinting . . . . .   175
　　6.6.4   Cell Viability Assays . . . . . . . . . . . . . . . . . . . . . . . . . . .   178

6.7     Advanced Extrusion-Based Bioprinting Techniques . . . . . . . . . . .   179
        6.7.1   Embedded Bioprinting . . . . . . . . . . . . . . . . . . . . . . .   179
        6.7.2   Multi-head Bioprinting . . . . . . . . . . . . . . . . . . . . . .   181
        6.7.3   Coaxial Bioprinting . . . . . . . . . . . . . . . . . . . . . . .   182
        6.7.4   Hybrid Bioprinting . . . . . . . . . . . . . . . . . . . . . . . .   183
6.8     Summary . . . . . . . . . . . . . . . . . . . . . . . . . . . . . . . . .   183
References . . . . . . . . . . . . . . . . . . . . . . . . . . . . . . . . . . . .   187

**7   Bioprinting of Vascular Networks in Scaffolds** . . . . . . . . . . . . . .   191
7.1     Introduction . . . . . . . . . . . . . . . . . . . . . . . . . . . . . . .   191
7.2     Blood Vessels and Formation Mechanisms . . . . . . . . . . . . . . . .   192
7.3     Bioinks for Vascular Networks . . . . . . . . . . . . . . . . . . . . . .   195
7.4     Bioprinting Vascular Networks . . . . . . . . . . . . . . . . . . . . . .   196
        7.4.1   Direct Bioprinting . . . . . . . . . . . . . . . . . . . . . . . .   196
        7.4.2   Indirect Bioprinting . . . . . . . . . . . . . . . . . . . . . . .   200
        7.4.3   Self-Assembled Vasculature Using Bioprinting . . . . . . . . .   204
7.5     Other Vascularization Approaches . . . . . . . . . . . . . . . . . . . .   207
7.6     Summary . . . . . . . . . . . . . . . . . . . . . . . . . . . . . . . . .   209
References . . . . . . . . . . . . . . . . . . . . . . . . . . . . . . . . . . . .   210

**8   Controlled Release of Biomolecules in Printed Scaffolds** . . . . . . . . .   213
8.1     Introduction . . . . . . . . . . . . . . . . . . . . . . . . . . . . . . .   213
8.2     Controlled Release of Biomolecules . . . . . . . . . . . . . . . . . . .   213
8.3     Biomolecules and Controlled Release Modes . . . . . . . . . . . . . .   215
        8.3.1   Biomolecules in Tissue Engineering . . . . . . . . . . . . . . .   215
        8.3.2   Modes of Controlled Release . . . . . . . . . . . . . . . . . .   216
8.4     Loading of Biomolecules into Scaffolds . . . . . . . . . . . . . . . . .   219
        8.4.1   Chemical Immobilization . . . . . . . . . . . . . . . . . . . . .   219
        8.4.2   Physical Encapsulation . . . . . . . . . . . . . . . . . . . . . .   220
        8.4.3   Micro/Nanoparticles . . . . . . . . . . . . . . . . . . . . . . .   221
8.5     Controlled Release via Micro/Nanoparticles . . . . . . . . . . . . . . .   222
        8.5.1   Techniques to Prepare Particles . . . . . . . . . . . . . . . . .   222
        8.5.2   Techniques to Characterize Controlled Release . . . . . . . . .   224
        8.5.3   Rate Programming of Particles . . . . . . . . . . . . . . . . . .   227
        8.5.4   Actively Controlled Release . . . . . . . . . . . . . . . . . . .   231
8.6     Design of a Controlled Release System . . . . . . . . . . . . . . . . . .   232
8.7     Summary . . . . . . . . . . . . . . . . . . . . . . . . . . . . . . . . .   233
References . . . . . . . . . . . . . . . . . . . . . . . . . . . . . . . . . . . .   235

**Index** . . . . . . . . . . . . . . . . . . . . . . . . . . . . . . . . . . . . .   237

# Chapter 1
# Extrusion Bioprinting of Scaffolds: An Introduction

## 1.1 Introduction

Millions of people suffer from tissue/organ injuries or damage, such as peripheral nerve injuries and heart attacks. Tissue/organ transplantation is the gold standard to treat some of these types of injuries but is severely restricted as an option due to the limited availability of donor tissues/organs. To address this issue, tissue engineering aims to produce tissue/organ substitutes or scaffolds to improve upon current treatment approaches, thus providing a permanent solution to damaged tissues/ organs. To this end, scaffolds made from biodegradable biomaterials play a crucial role in supporting/promoting cell growth and tissue regeneration as well as transporting nutrients and wastes. This chapter presents a brief introduction to the development of scaffolds for tissue engineering and modeling as well as various scaffold fabrication techniques, including extrusion bioprinting.

## 1.2 Scaffolds for Tissue Engineering and Modeling

Tissue engineering (TE) aims to produce tissue/organ substitutes or scaffolds to be implanted into human patients to improve upon current treatment approaches, thus providing a permanent solution to damaged tissues/organs. An analogy would be buying new parts at the mechanic to replace car parts that are broken or no longer functioning. Success in tissue engineering would mean that someone who unfortunately suffers a tissue/organ injury could go to a hospital, have the engineered substitute implanted into his/her body, and later completely recover the function of a healthy body with the help of the engineered substitute.

The general principle behind TE is schematically shown in Fig. 1.1. Cells from a patient (or other resource) are harvested and then seeded onto or incorporated into an engineered substitute or scaffold (typically along with growth factors or other



D. X. B. Chen, *Extrusion Bioprinting of Scaffolds for Tissue Engineering*, https://doi.org/10.1007/978-3-031-72471-8_1

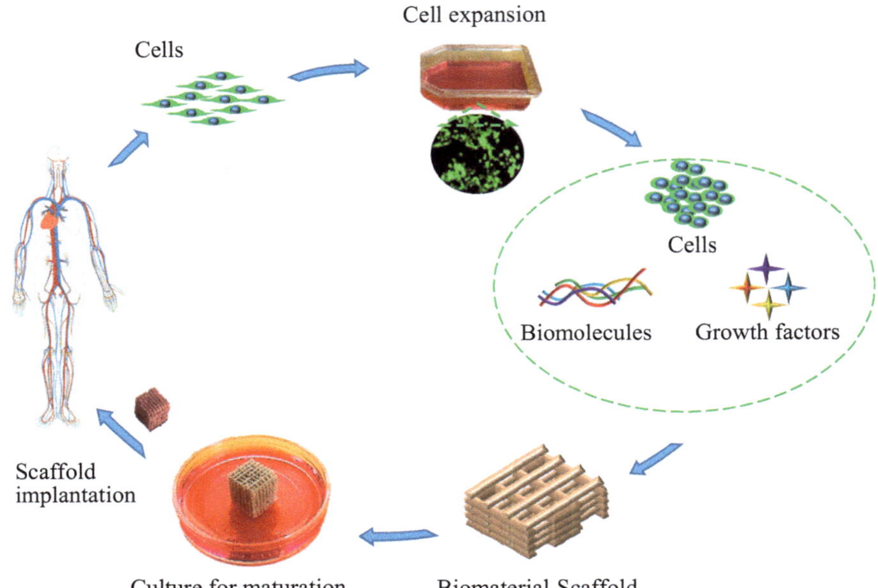



**Fig. 1.1**  General principle behind tissue engineering

biomolecules to stimulate cell growth and functions). The cell-incorporated scaffold is then cultured to maturation, resulting in a functional construct that is then implanted into the patient to help repair or heal the damaged tissue/organ. Scaffold-based TE is an interdisciplinary field that involves applying the principles of life sciences and engineering to repair damaged tissues and organs with the help of scaffolds [1].

Made from biodegradable biomaterials (such as polymers), TE scaffolds are used to support and facilitate cell/tissue growth and transport of nutrients and wastes, while degrading gradually during the healing process. Scaffolds are also created to mimic human tissue/organs (e.g., human lung), serving as in vitro (out of body) tissue or virus-disease models to test/validate newly developed therapeutics and vaccines prior to their use in humans [2, 3]. Several functional requirements have been identified as crucial for these scaffolds in terms of architectural, mechanical, and biological properties.

*Architectural properties* of a scaffold refer to its external geometry and internal structure. Generally speaking, a scaffold's external geometry should mimic that of the tissue/organ to be repaired, while its internal structure should be highly porous to allow for cell growth and movement as well as facilitate the transport of nutrients into the scaffold and the removal of metabolic wastes out of the scaffold during the healing process.

*Mechanical properties* of a scaffold refer to its mechanical strength and degradation. During the healing process, the scaffold materials degrade as the cells/tissue grows and, as a result, the mechanical strength of the scaffold decreases with time. Concurrently, the cells grow and the tissue regenerates, which imparts mechanical strength to the combined construct of scaffold material and regenerated tissue. It is generally accepted that the mechanical strength of a scaffold at the initial stage of implantation or of a combined construct of scaffold and regenerated tissue during the healing process should be similar to that of the tissue/organ being repaired.

*Biological properties* of a scaffold refer to its ability to support cell growth/ functions (such as cell attachment, proliferation, and differentiation) and tissue regeneration, with limited or no negative effects (such as inflammation) on the host system (i.e., animal or human) in which the tissues/organs are to be repaired. The biological properties of a scaffold are typically evaluated using in vitro and in vivo tests. In vitro (literally "in glass") tests take place in a well-controlled laboratory environment, while in vivo tests are performed in the living body of an animal or human.

Depending on the TE applications or the tissues/organs to be repaired or mimicked, more requirements may be imposed on the scaffolds. For example, scaffolds for peripheral nerve repair should possess a biodegradable and porous channel wall and incorporate viable Schwann cells [4, 5], which facilitate axon growth and thus functional recovery.

The development of TE scaffolds consists of three stages—design, fabrication, and characterization—as shown in Fig. 1.2. Based on the functional requirements, TE scaffolds should generally be designed with three-dimensional (3D) and porous structures of appropriate mechanical and biological properties, where the key is to design and/or select the scaffold biomaterials, internal structure, and living cells to be seeded on or incorporated within the scaffolds. Typically, scaffold design starts from an understanding and/or knowledge of the architecture of the tissue/organ to be

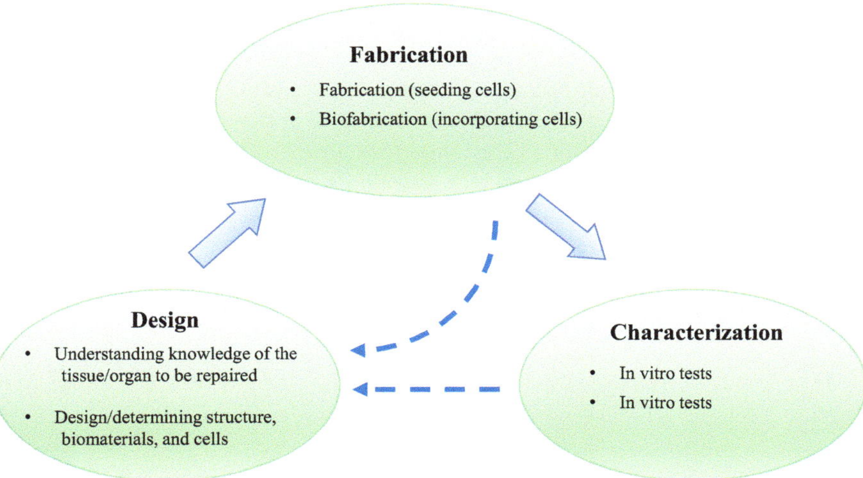


**Fig. 1.2** Schematic of the development of TE scaffolds

repaired; medical imaging technologies, such as computed tomography and magnetic resonance imaging, are common tools for this purpose [6, 7]. With such knowledge, scaffolds are designed with appropriate external geometries and internal structures, as well as specifically chosen and spatially arranged biomaterials/cells, so as to mimic the architectural, mechanical, and biological properties of the tissues/organs to be repaired.

In the second stage of scaffold development, scaffolds are created from biomaterials and living cells, as designed, by means of fabrication techniques. Scaffolds can be either fabricated from biomaterials and subsequently seeded with living cells or fabricated from biomaterials incorporating living cells (known as biofabrication). Seeding cells onto scaffolds after their fabrication imposes limits on the ability to spatially place living cells into scaffolds as well as on seeding depth, that is, the cells seeded into the scaffold remain near the scaffold surface. Advantages of incorporating cells in the biofabrication process include the ability to produce a spatial distribution of cells, thus allowing the cell organization of the target tissue/organ to be mimicked. Sustaining the viability of living cells during the fabrication process is essential, and emphasizes the importance of sterile and gentle conditions for scaffold biofabrication.

The last stage of scaffold development is scaffold characterization. By means of in vitro and in vivo tests, the performance or outcomes of scaffolds are examined and analyzed in terms of architectural, mechanical, and biological properties for various TE applications. In many cases, the scaffolds, once fabricated, need to be cultured in vitro prior to their implantation to facilitate their maturation for optimal in vivo performance or outcomes.

Figure 1.2 shows the development of TE scaffolds as continuous and cyclic in nature. Scaffold development is not linear; that is, one does not necessarily achieve the best scaffold by simply proceeding from one stage to the next. The development of a scaffold for a given tissue engineering application is typically accomplished by iteration through the aforementioned three stages. For example, new discoveries in the relationship between the function and structure of tissues/organs help and improve our understanding of the architecture of tissues/organs to be repaired and therefore the scaffold design. Scaffolds should also be designed such that they can be fabricated by means of existing fabrication techniques, while advances in scaffold fabrication allow for improvement over existing scaffold designs and more functional scaffolds. The performance and/or outcomes of scaffolds, as examined in vitro and in vivo, not only illustrate the effectiveness of the scaffold design and fabrication but also provide a means of feedback to refine the scaffold design, as well as advance fabrication techniques to achieve better outcomes for TE applications.

## 1.3   Scaffold Fabrication

A number of fabrication techniques have been applied to fabricate scaffolds from biomaterials. Generally, these techniques are divided into three categories, i.e., traditional, electrospinning, and 3D printing.

### *1.3.1   Traditional Techniques*

Traditional techniques refer to those that are adopted from traditional fields to process biomaterials into scaffolds with a randomly generated pore structure. These techniques include porogen leaching, gas foaming, phase separation, melt molding, and freeze drying.

**Porogen Leaching** Porogen leaching is one of the oldest polymer processing techniques to make porous products and, in the early days of TE, was widely used to fabricate scaffolds. This technique involves dispersing a template (e.g., salt particles) within a polymer solution, gelling or fixing the template/polymer structure, and then removing or leaching the template from the structure so as to create a scaffold with a porous structure (Fig. 1.3a).

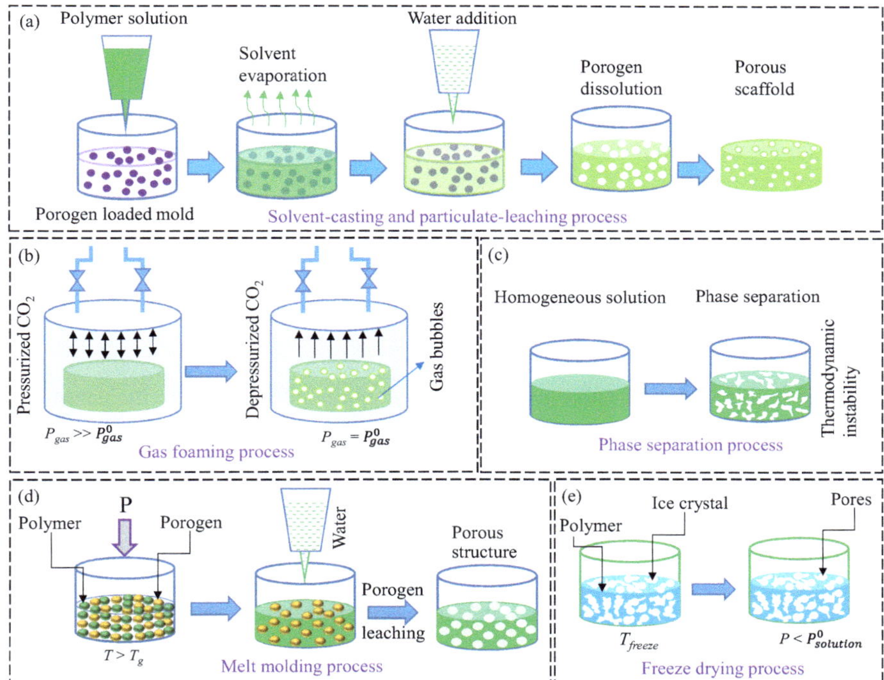


**Fig. 1.3** Schematic of conventional scaffold fabrication techniques: (**a**) solvent casting and porogen leaching, (**b**) gas foaming, (**c**) phase separation, (**d**) melt molding, and (**e**) freeze drying

**Gas Foaming** During the gas-foaming process (Fig. 1.3b), molded polymers are pressurized with gas-foaming agents, such as $CO_2$ and nitrogen; the release of pressure then results in the nucleation and growth of gas bubbles and thus porous scaffold structures. This technique has the advantage of being an organic solvent-free process for scaffold fabrication; the major drawback is that the process may yield structures with largely unconnected pores and a non-porous external surface.

**Phase Separation** During the phase separation process (Fig 1.3c), a polymer solution is quenched and undergoes a liquid-liquid phase separation to form two phases: a polymer-rich phase and a polymer-poor phase. The polymer-rich phase solidifies and the polymer poor phase is removed, leaving a highly porous polymer network. The micro- and macro-structures of the resulting scaffold are controlled by varying process parameters, such as polymer concentration, quenching temperature, and quenching rate. The process can be conducted at low temperatures, which is beneficial for the incorporation of bioactive molecules in the structure.

**Melt Molding** During the melt molding process (Fig. 1.3d), a mold is filled with polymer powder and a porogen component and then heated to above the glass-transition temperature of the polymer ($T_g$), causing the materials to bind together to form a scaffold in the shape of the mold. The porogen is then leached out, leaving a scaffold with a porous structure. Melt molding with porogen leaching is a nonsolvent fabrication process that allows independent control of morphology and shape. Drawbacks include the possibility of residual porogen and high processing temperatures that preclude the ability to incorporate bioactive molecules.

**Freeze Drying** During the freeze-drying process (Fig. 1.3e), a polymer solution is cooled to the temperature at which all materials become solid; the solvent is then sublimed from the solid phase to the gas phase by reducing the pressure to below the equilibrium vapor pressure of the frozen solvent. By doing so, the solvent is removed, leaving a scaffold with a porous structure. The scaffold structure depends on the concentration of the polymer solution, freezing rate, and applied pressure.

## 1.3.2  Electrospinning

*Electrospinning* is a fabrication technique to create fine fibers down to the nanometer scale from polymer solutions or melts. This technique was first developed in the 1930s, and since 1990 has found widespread applications in the fabrication of TE scaffolds. A typical electrospinning setup, as schematically shown in Fig. 1.4, includes three basic components: a spinneret (or a small orifice and flat-tipped needle), a voltage source, and a collector. During scaffold fabrication, a high voltage is applied to the polymer solution in the spinneret, while the collector is grounded; as a result, a large electric field is generated between the polymer solution and collector that causes the polymer solution to be continuously ejected from the spinneret. The jet travels spirally and lands on the collectors, forming a 3D scaffold of fibrous

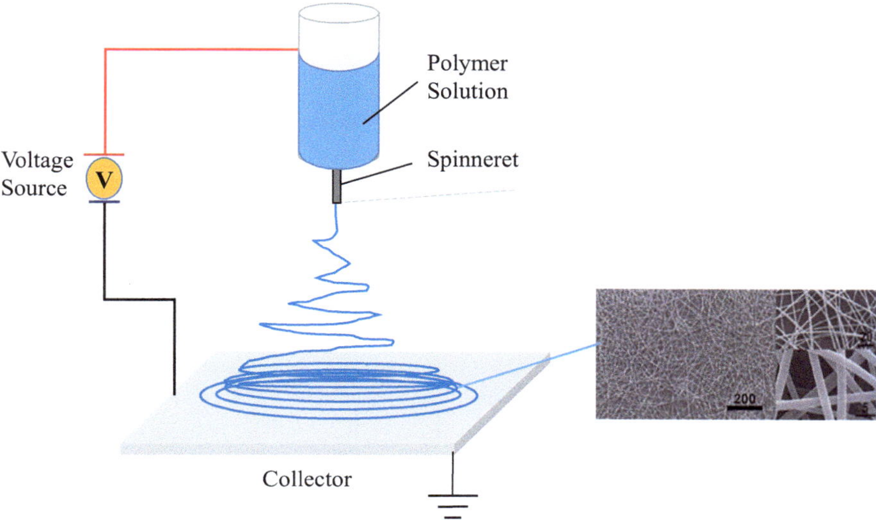

**Fig. 1.4**  Working principle of electrospinning

architecture. Depending on the process parameters for spinning (e.g., the applied voltage and the distance between the spinneret and collector), the diameter of spun fibers typically varies between 200 nm and 5 μm.

With current advances, spinnerets can be designed to deliver multiple polymer solutions. For example, a coaxial spinneret with an inner needle and an external needle can be used to apply two polymer solutions, respectively, forming fibers with a core/shell structure. Cells can also be added to the electrospinning solution to form cell-incorporated scaffolds. In cases in which solvent accumulation or toxicity is a concern, electrospinning polymers without solvents (via melting), called *melt electrospinning*, can be used to create scaffolds.

## *1.3.3   Three-Dimensional Printing*

Three-dimensional or 3D printing of scaffolds refers to the technique of depositing or printing biomaterials in a layer-by-layer manner to create scaffolds with a 3D structure. Distinctive from the aforementioned traditional techniques, 3D printing offers reproducible control over the architectural properties of scaffolds due to the multilayer deposition of biomaterials. Other merits include the process being: (1) easy and straightforward for the creation of scaffolds with porous structures, (2) able to create complex structures to mimic those of natural tissues/organs, and (3) capable of incorporating living cells during scaffold fabrication. Based on the working principles, techniques used for 3D printing can be classified as extrusion, ink-jet, or laser-assisted.

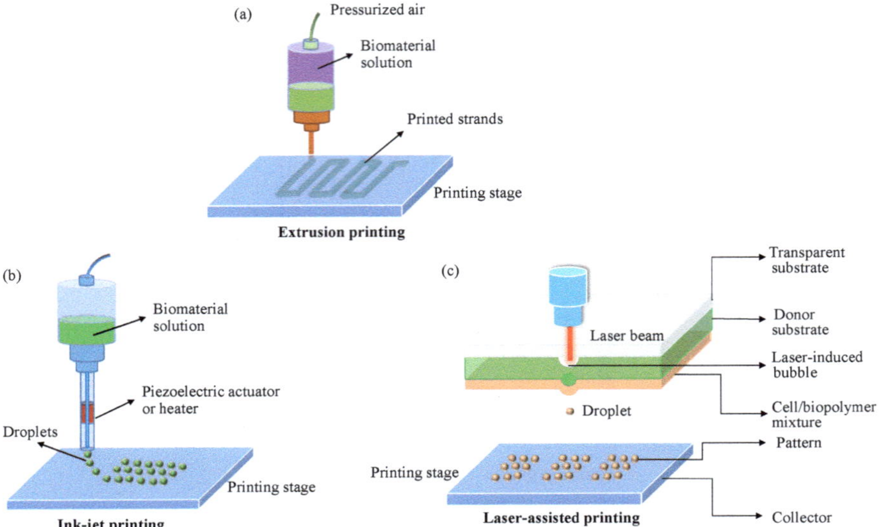

**Fig. 1.5** 3D printing techniques: (**a**) extrusion printing, (**b**) ink-jet printing, and (**c**) laser-assisted printing

**Extrusion Printing** Extrusion printing is a technique to extrude or dispense continuous strands or fibers of biomaterials, layer-by-layer, to form 3D structures [7–9]. Extrusion printing is based on the principle of fluid extrusion or dispensing (Fig. 1.5a), where the biomaterial solution stored in a syringe is driven by mechanical force (e.g., pressurized air) through a needle and then onto a printing stage.

**Ink-Jet Printing** Adopted from the working principle of a commercial printer, ink-jet printing propels droplets of biomaterial solution (or the ink in a printer) onto a printing stage (Fig. 1.5b). As such, ink-jet printing is also known as drop-on-demand printing. The forces to propel the droplets of solution can be generated thermally or acoustically.

**Laser-Assisted Printing** Laser-assisted printing is based on the principle of laser-induced forward transfer (Fig. 1.5c); when the laser pulses focus and hit biomaterials covered in an energy-absorbing substrate, high pressures are generated that propel the biomaterials onto a collector substrate. Laser-assisted printing is performed without the need for needles, thus avoiding the issue of clogging, which can occur with other printing techniques.

## 1.4  Extrusion Bioprinting of Scaffolds

By 3D extrusion printing, scaffolds can be fabricated from biomaterials mixed with living cells. This fabrication process is referred to as *extrusion bioprinting*, and the biomaterial solution mixed with living cells is referred to as the *bioink* (similar to "ink" for a printer). An extrusion bioprinting system typically consists of a printing head, a positioning-control component, and a temperature-control component. A schematic of such a system is shown in Fig. 1.6a. The system includes a dispenser or printer mounted on the printing head, which can be controlled to move in three directions, a printing/supporting stage or platform to support the scaffold being fabricated, and three controllers that control printing, positioning, and temperature.

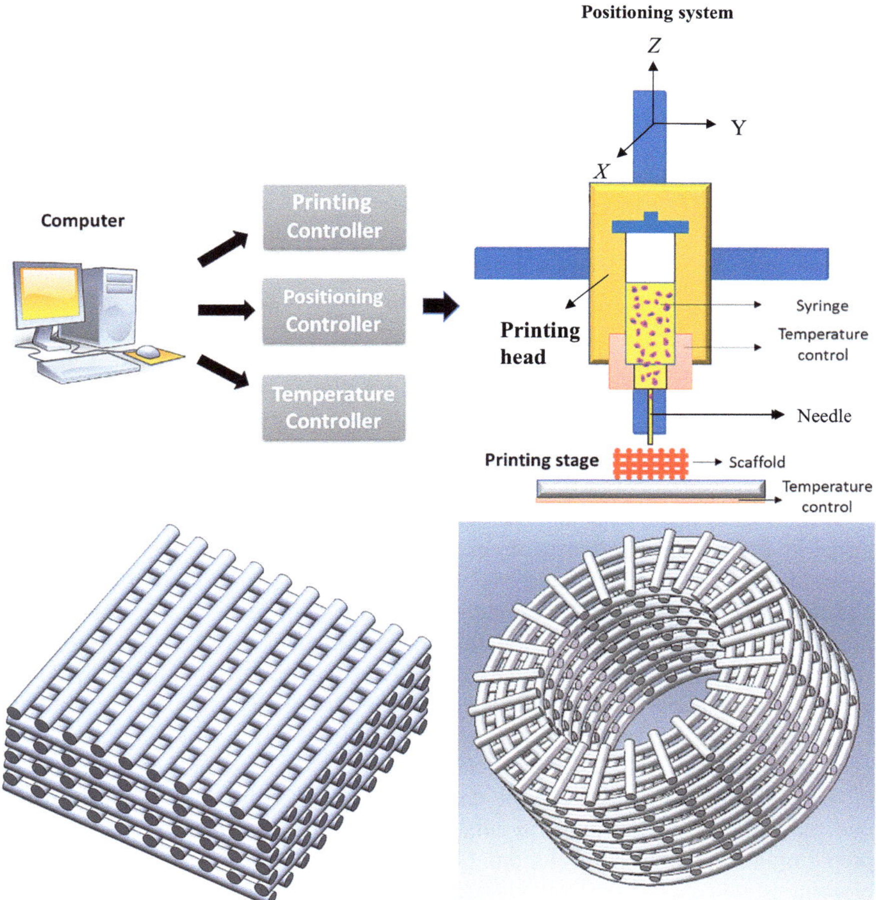


**Fig. 1.6**  Schematic of (**a**) extrusion-based bioprinting and (**b**) typical structures of printed scaffolds

During bioprinting, the bioink is loaded into the syringe on the printing head and then driven by mechanical force (e.g., pressurized air) through a needle onto the printing stage, forming a layer-structure scaffold. Depending on the internal diameter of the needle used for printing, the resolution of strands that can be achieved is on the order of 100–150 μm. Typical scaffolds fabricated by such systems have 3D structures with repeatable layers of printed strands, as shown in Fig. 1.6b).

Bioprinting allows for the incorporation of living cells within scaffolds. Notably, living cells are dynamic structures with functions (e.g., growth and proliferation) that may be degraded by mechanical forces. During the bioprinting process, cells are subjected to sustained process-induced forces, such as pressure, shear stress, and extensional stress, which can cause the deformation and breach of cell membranes. Although cells have elastic abilities to resist a certain level of mechanical force, cell membranes may lose their integrity if the applied force exceeds a certain threshold; as a result, cells may be damaged and even lose their functions and viability [10].

To preserve *cell viability*, the solution used for bioprinting must be bio-compatible while allowing rapid transport of nutrients/metabolites to/from incorporated cells. Hydrogels have been widely used for cell incorporation in bioprinting. A hydrogel is a gelled or cross-linked (via either physical or chemical bonding) network of polymers, such as collagen, alginate, chitosan, or polylactic acid. The cross-linked network possesses high water content among polymer chains, which if used for cell incorporation provide a hydrated tissue-like environment, thus enhancing the cell viability in bioprinting. The crossed-linked network greatly facilitates the formation of a 3D structure when printing scaffolds. The gelation or crossing-linking of hydrogels takes time, and during this gelation period, the hydogel is in a solution or semi-solution form and is able to flow or spread on the printing stage. As a result, the printed structure of a scaffold may not be the same as the one designed. In some cases, the printed structures even collapse and fail to form a 3D structure; such hydrogels would be deemed unprintable. Examination of the difference between the scaffold design and the printed structures is a common practice to measure *printability* in bioprinting [11, 12].

A number of factors can be involved in the bioprinting process to determine and affect its performance in terms of cell viability and printability. These factors are generally associated with biomaterial solution properties, the bioprinting process, and the scaffold design [12]. Biomaterial solution properties include the physical properties (such as contact angle), flow behavior, and crosslinking mechanisms, while the printing process involves the mechanical force (e.g., pressurized air) applied for printing, printing-head movement speed, and structural parameters (e.g., needle diameter and length). Notably, scaffold design in terms of, for example, geometry/structure, can also be relevant and affect the performance of the bioprinting process. To achieve the desired or optimized bioprinting performance, one may rigorously design or regulate (1) biomaterial-solution properties when preparing the biomaterial solution, (2) printing conditions when designing the bioprinting process, and (3) scaffold geometry and structures.

## 1.5  Advances and Limitations of Extrusion Bioprinting

Over the last two decades, advances in both engineering techniques and life sciences have evolved extrusion bioprinting from a simple technique to one able to create diverse, yet complicated, tissue scaffolds from a wide range of biomaterials and cell types. These scaffolds have been used in various tissue engineering and modeling applications, including the repair of damaged skin [13], cartilage [8], bones [14], nerves [4], teeth [15], and spinal cords [16]; treatment of corneal blindness [17] and heart attack [18]; and lung modeling [3] and testis modeling [19].

Extrusion bioprinting allows for the printing of a wide array of biomaterials, which are typically polymers. Polymers are organic biomaterials possessing long chains with high water contents and thus are able to provide a hydrated tissue-like environment that supports cell functions including cell attachment, proliferation, and differentiation, as well as tissue regeneration. Most polymers used for bioprinting are derived from natural extracellular matrix sources, including collagen, gelatin, fibrin, hyaluronic acid, alginate, agarose, and decellularized extracellular matrix (dECM) [20, 21]. These naturally derived polymers have intrinsic capabilities to support cell viability and proliferation and can be relatively easily degraded or metabolized. However, limitations of natural biomaterials include rejection and/or immune-related sequelae, quick degradation rates, and poor mechanical properties. Synthetic polymers, such as polycaprolactone (PCL) and polyethylene glycol (PEG), are usually biologically inert but exhibit strong and robust mechanical properties. While materials science continues to develop and synthesize new polymers with more appropriate properties for extrusion bioprinting, research has also begun to use two or more polymers or composite polymers to formulate solutions or bioinks. Composites with the incorporation of inorganic fillers for improved mechanical properties, electrical properties, and biological properties, have also been drawing considerable attention for future advances.

Extrusion bioprinting can produce tissue scaffolds using various cells types, including both primary cells and stem cells. Primary cell types, which are isolated from animals or humans and include cells such as osteocytes, chondrocytes, and keratinocytes, have been used in tissue scaffolds to faithfully represent tissues including bones, cartilage, and skin, respectively. In some cases, primary cells isolated from living tissues may be difficult or challenging to culture. In such cases, stem cells are often used as a substitute for primary cells in tissue scaffold bioprinting. Stem cells can self-renew and differentiate into specific cell types when certain cues are provided. The extrusion bioprinting technique has shown great potential for regulating and conducting stem cell growth and differentiation in many applications, such as those targeting brain tissue, gingival tissue, adipose tissue, and bone marrow tissue. Bioprinting scaffolds with multiple materials and cells have also attracted considerable attention [22], along with the help of machine learning [23]. Various extrusion-based bioprinting techniques have been developed and advanced, such as embedded bioprinting, multi-head bioprinting, co-axial bioprinting, and hybrid bioprinting [20]. More advanced extrusion-based bioprinting techniques are expected to be developed in the future to create scaffolds that better mimic tissue structures.

Various printed structures such as beads, filaments, fibers, channels, sheets, rolls, grids, and porous 3D constructs that mimic various tissue components have been successfully printed at the micro or macro scale. Among these structures, the formation of vasculature is one of the major challenges for tissue engineering. Notably, vasculature is essential to maintain the viability and biological function of large cell populations in growing tissue. In vivo, well-distributed vascular capillaries are seen in different tissues at a distance of every ~100–200 μm [24]. Similarly, tissue regeneration with the aid of scaffolds, particularly large and thick scaffolds, requires the incorporation of an interconnected vascular network to facilitate mass transfer of nutrients, signaling molecules, oxygen, growth factors, metabolic waste, etc. between the cells in the scaffold and the culture medium [9, 24]. In extrusion-based bioprinting, vessel-like permeable channels have been produced and used to facilitate vascularization with the expectation of forming vascular networks. Supporting cells such as endothelial cells are often deposited in vessel-like channels during bioprinting to initiate the formation of vasculature and subsequently support their stabilization and function, which can further facilitate the angiogenesis of vessel networks [24].

Extrusion bioprinting has many advantages over ink-jet and laser-assisted printing. Extrusion bioprinting allows for a wide array of biomaterials and cells to be printed, including both native and synthetic hydrogel polymers, cell aggregates, and dECM. Also, extrusion bioprinting can facilitate the deposition of biomaterials with physiological cell density, which is a major challenge with other bioprinting techniques. Due to its fast deposition speed, extrusion bioprinting has often been used to produce large-scale scaffolds. However, extrusion bioprinting also has several limitations. It has a limited strand resolution (typically greater than 100 μm), mainly due to the issues associated the mechanical force required to drive a scaffold solution through the fine needle as well as the process-induced forces on the cells incorporated. In contrast, laser-assisted bioprinting techniques are capable of reaching nanometer resolutions and inkjet-based bioprinting can form droplets in a diameter down to a few μm. In addition, the printability of hydrogels by extrusion bioprinting is heavily dependent on their crosslinking capability and/or the printing conditions; biomaterials with a slow crosslinking speed may not be appropriate for use in bioprinting due to difficulties related to the formation of 3D structures. In addition, needle clogging with biomaterial solution is another problem in extrusion bioprinting that may cause complete interruption of biomaterial deposition and therefore affect the integrity of the resulting scaffold structure.


## 1.6  Summary

Tissue engineering aims to produce tissue/organ substitutes that improve upon current treatment approaches, thus providing a permanent solution to damaged tissues/organs. In scaffold-based TE, scaffolds are used to support and facilitate cell/tissue growth and the transport of nutrients and wastes, while gradually degrading during the healing process. The scaffolds can also be created as tissue or virus-disease models for testing or validating newly developed therapeutics and vaccines prior to their use in humans.

Several requirements have been identified as crucial for TE scaffolds in terms of architectural, mechanical, and biological properties. The architectural properties of a scaffold are characterized by its external geometry and internal structure, mechanical properties by its mechanical strength and degradation, and biological properties by its ability to support cell growth/functions and tissue regeneration with limited or no negative effects on the host system. The biological properties of a scaffold are typically evaluated using in vitro and in vivo tests.

TE scaffolds can be fabricated by conventional, electrospinning, and 3D printing techniques. Traditional techniques are those adopted from traditional fields to process biomaterials into scaffolds with a randomly generated pore structure. These techniques include porogen leaching, gas foaming, phase separation, melt molding, and freeze drying. Electrospinning is a fabrication technique to create fine fibers up to the nanometer scale from polymer solutions or melts. 3D printing refers to extrusion, ink-jet, and laser-assisted printing techniques that are able to deposit or pattern biomaterials in a layer-by-layer manner to create scaffolds or constructs with a 3D structure. Distinct from traditional techniques, 3D printing offers reproducible control over the architectural properties of scaffolds. By bioprinting, scaffolds can be fabricated from biomaterials mixed with living cells (or bioink). Cell viability and printability are two aspects of the performance of the bioprinting process that are determined and affected by biomaterial-solution properties, printing conditions, and scaffold design.

Extrusion bioprinting can fabricate tissue scaffolds with various structures by incorporating primary cells and/or stem cells. These fabricated scaffolds have been widely used in TE and modeling applications including the repair of damaged skin and cartilage, treatment of heart attack, and lung modeling. Extrusion bioprinting has numerous merits and demerits compared to other printing techniques.

**Problems**
1. Explain the general principle used in scaffold-based tissue engineering to heal damaged tissues/organs and the role the scaffold plays in the healing process.
2. Name the three requirements imposed on TE scaffolds. Perform a literature review on one requirement and illustrate your understanding of this requirement.
3. Name the techniques that can be used to fabricate tissue scaffolds. Perform a literature review on one technique (except extrusion bioprinting) and explain its working principle and its merits/demerits for use in scaffold fabrication.
4. Briefly explain the process of extrusion bioprinting of scaffolds, along with its two key issues, that is, printability and cell viability.
5. Name and explain one achievement accomplished by means of extrusion bioprinting that has been reported in the literature.
6. Name and explain one advantage and one disadvantage of extrusion bioprinting compared to ink-jet and laser-assisted printing.

# References


1. R. Langer, J.P. Vacanti, Tissue engineering. Science **260**, 920–926 (1993). https://doi.org/10.1126/science.8493529
2. A. Zimmerling, X. Chen, Bioprinting for combating infectious diseases. Bioprinting **20**, e00104 (2020). https://doi.org/10.1016/j.bprint.2020.e00104
3. A. Zimmerling, N.A. Dahlan, Y. Zhou, X. Chen, Recent frontiers in biofabrication for respiratory tissue engineering. Bioprinting **40**, e00342 (2024). https://doi.org/10.1016/j.bprint.2024.e00342
4. M.D. Sarker, S. Naghieh, A.D. McInnes, et al., Strategic design and fabrication of nerve guidance conduits for peripheral nerve regeneration. Biotechnol. J. **13**, 1700635 (2018). https://doi.org/10.1002/biot.201700635
5. A. Rajaram, X.B. Chen, D.J. Schreyer, Strategic design and recent fabrication techniques for bioengineered tissue scaffolds to improve peripheral nerve regeneration. Tissue Eng. Part B Rev. **18**, 454–467 (2012). https://doi.org/10.1089/ten.teb.2012.0006
6. S.V. Murphy, A. Atala, 3D bioprinting of tissues and organs. Nat. Biotechnol. **32**, 773–785 (2014). https://doi.org/10.1038/nbt.2958
7. L. Ning, X. Chen, A brief review of extrusion-based tissue scaffold bio-printing. Biotechnol. J. **12**, 1600671 (2017). https://doi.org/10.1002/biot.201600671
8. F. You, B.F. Eames, X. Chen, Application of extrusion-based hydrogel bioprinting for cartilage tissue engineering. Int. J. Mol. Sci. **18**, 1597 (2017). https://doi.org/10.3390/ijms18071597
9. S. Naghieh, M.D. Sarker, M. Izadifar, X. Chen, Dispensing-based bioprinting of mechanically-functional hybrid scaffolds with vessel-like channels for tissue engineering applications – a brief review. J. Mech. Behav. Biomed. Mater. **78**, 298–314 (2018). https://doi.org/10.1016/j.jmbbm.2017.11.037
10. M. Li, X. Tian, N. Zhu, et al., Modeling process-induced cell damage in the biodispensing process. Tissue Eng. Part C Methods **16**, 533–542 (2010). https://doi.org/10.1089/ten.tec.2009.0178
11. S. Naghieh, X. Chen, Printability–a key issue in extrusion-based bioprinting. J. Pharm. Anal. **11**, 564–579 (2021). https://doi.org/10.1016/j.jpha.2021.02.001
12. Z. Fu, S. Naghieh, C. Xu, et al., Printability in extrusion bioprinting. Biofabrication **13**, 033001 (2021). https://doi.org/10.1088/1758-5090/abe7ab
13. L.Y. Daikuara, X. Chen, Z. Yue, et al., 3D bioprinting constructs to facilitate skin regeneration. Adv. Funct. Mater. **32**, 2105080 (2022). https://doi.org/10.1002/adfm.202105080
14. Z. Yazdanpanah, J.D. Johnston, D.M.L. Cooper, X. Chen, 3D bioprinted scaffolds for bone tissue engineering: State-of-the-art and emerging technologies. Front. Bioeng. Biotechnol. **10**, 824156 (2022). https://doi.org/10.3389/fbioe.2022.824156
15. F. Mohabatpour, X. Chen, S. Papagerakis, P. Papagerakis, Novel trends, challenges and new perspectives for enamel repair and regeneration to treat dental defects. Biomater. Sci. **10**, 3062–3087 (2022). https://doi.org/10.1039/D2BM00072E
16. M. Wang, P. Zhai, X. Chen, et al., Bioengineered scaffolds for spinal cord repair. Tissue Eng. Part B Rev. **17**, 177–194 (2011). https://doi.org/10.1089/ten.teb.2010.0648
17. R. Wang, S. Deng, Y. Wu, et al., Remodelling 3D printed GelMA-HA corneal scaffolds by cornea stromal cells. Colloids Interface Sci. Commun. **49**, 100632 (2022). https://doi.org/10.1016/j.colcom.2022.100632
18. F. Ketabat, J. Alcorn, M.E. Kelly, et al., Cardiac tissue engineering: A journey from scaffold fabrication to in vitro characterization. Small Sci. (2024). https://doi.org/10.1002/smsc.202400079
19. T.C. Cham, X. Chen, A. Honaramooz, Current progress, challenges, and future prospects of testis organoids. Biol. Reprod. **104**, 942–961 (2021). https://doi.org/10.1093/biolre/ioab014
20. X.B. Chen, A. Fazel Anvari-Yazdi, X. Duan, et al., Biomaterials/bioinks and extrusion bioprinting. Bioact. Mater. **28**, 511–536 (2023). https://doi.org/10.1016/j.bioactmat.2023.06.006

21. A.D. McInnes, M.A.J. Moser, X. Chen, Preparation and use of decellularized extracellular matrix for tissue engineering. J. Funct. Biomater. **13**, 240 (2022). https://doi.org/10.3390/jfb13040240
22. N. Betancourt, X. Chen, Review of extrusion-based multi-material bioprinting processes. Bioprinting **25**, e00189 (2022). https://doi.org/10.1016/j.bprint.2021.e00189
23. A. Malekpour, X. Chen, Printability and cell viability in extrusion-based bioprinting from experimental, computational, and machine learning views. J. Funct. Biomater. **13**, 40 (2022). https://doi.org/10.3390/jfb13020040
24. M.D. Sarker, S. Naghieh, N.K. Sharma, et al., Bioprinting of vascularized tissue scaffolds: Influence of biopolymer, cells, growth factors, and gene delivery. J. Healthc. Eng. **2019**, 9156921 (2019). https://doi.org/10.1155/2019/9156921

# Chapter 2
# Tissue Scaffold Design

## 2.1  Introduction

In tissue engineering or modeling, scaffolds support and facilitate cell/tissue growth, transport nutrients and wastes, and gradually degrade during the repair process of damaged tissue/organs. The crucial requirements to achieve this functionality can be classified into three categories: architectural, mechanical, and biological. Based on these requirements, scaffold design typically follows a process that involves an understanding of the tissues/organs to be repaired, determination of the required architecture or structure of the scaffold, and selection of living cells and biomaterials to fabricate the scaffold. This chapter introduces these crucial requirements imposed on scaffolds and, on this basis, describes the scaffold design process along with key design parameters. Three scaffold structures, i.e., mono, hybrid, and zonal structures, that can be fabricated using current bioprinting techniques are also discussed with respect to scaffold design.

## 2.2  General Requirements of Tissue Scaffolds

### 2.2.1  Architectural Properties

The architectural properties of scaffolds include the external geometry and internal structure. The external geometry of a scaffold should be appropriate for its implantation into the injured site, while the internal structure should be highly porous with interconnected pores. Such a porous structure allows for not only cell growth and movement but also nutrient transport into the scaffold and metabolic waste removal from the scaffold during the repair process. Pore size and porosity are commonly used parameters to characterize the porous structure.



D. X. B. Chen, *Extrusion Bioprinting of Scaffolds for Tissue Engineering*,
https://doi.org/10.1007/978-3-031-72471-8_2

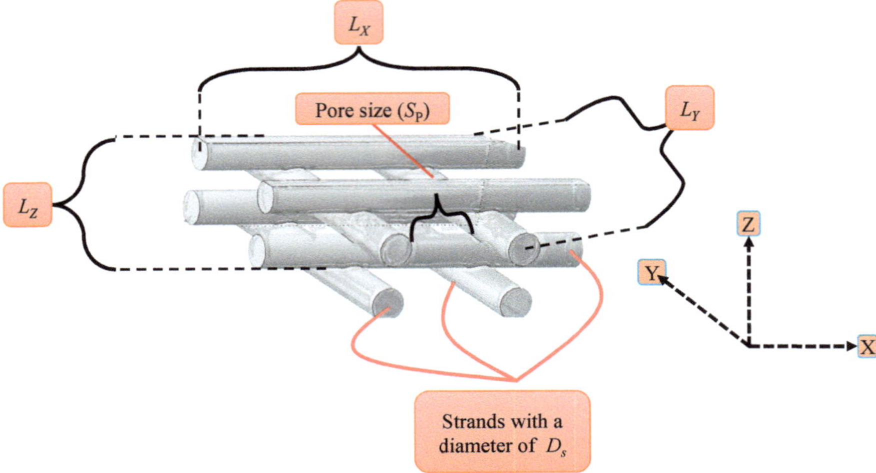


**Fig. 2.1** Architecture of a scaffold fabricated by an extrusion-based bioprinting technique

Figure 2.1 illustrates the architecture of a cubic scaffold, where the fibers or strands are cylindrical in shape and evenly arranged within the scaffold. In Fig. 2.1, $D_s$ denotes the strand diameter and $L_x$, $L_y$, and $L_z$ denote the scaffold dimensions in the $X$, $Y$, and $Z$ directions, respectively. The pore size ($S_p$) in each of the $X$, $Y$, and $Z$ directions is the distance between two adjacent, yet parallel, strands in that direction, and the porosity is the void or empty volume ($V_{\text{void}}$) expressed as a percentage of the total volume ($V_{\text{total}}$) of the scaffold, i.e.,

$$\%\text{Porosity} = \frac{V_{\text{void}}}{V_{\text{total}}} \times 100\% \tag{2.1}$$

In the above equation, $V_{\text{total}}$ includes both the void volume and the scaffold material volume and is given by $L_x \times L_y \times L_z$. In an ideal situation where the pore size in the vertical or $Z$ direction is equal to the strand diameter, the porosity of a cubic scaffold, as defined in Eq. (2.1), can then be rewritten as:

$$\%\text{Porosity} = \frac{V_{\text{total}} - \sum \text{Volume of all strands}}{V_{\text{total}}} \times 100\%$$
$$= \frac{L_x \times L_y \times L_z - \left( N_1 \times L_x + N_2 \times L_y \right) \times \pi \times \left(\frac{D_s}{2}\right)^2}{L_x \times L_y \times L_z} \times 100\%, \tag{2.2}$$

where $N_1$ and $N_2$ are the number of strands of all layers parallel to the $X$ and $Y$ directions, respectively. Notably, due to bioprinting imperfections or printability (as discussed in Chap. 3), the bioprinted scaffold dimensions may be different from those as designed. In such cases, direct measurement or characterization of strand

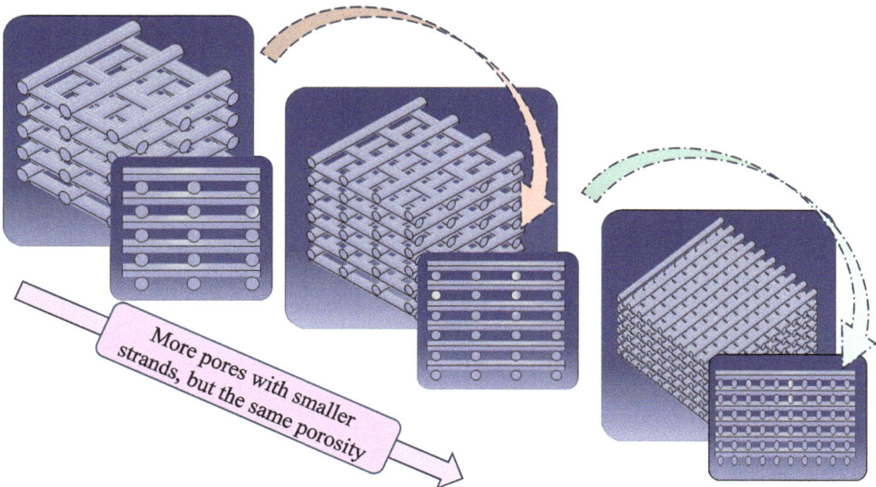



**Fig. 2.2**   Scaffolds with varying strand diameters and pore sizes, but with the same porosity

diameter and porosity by means of various techniques is typically conducted, as discussed in Chap. 6.

Equation 2.2 shows that scaffold porosity is determined by the strand diameter and the number of strands in each layer. Changing the strand diameter and the number of strands in each layer of a scaffold can be done to achieve a given porosity, as illustrated in Fig. 2.2. Notably, these scaffolds in Fig. 2.2 have different surface areas, which may lead to different behavior in terms of transporting nutrients and wastes [1] as well as varying degradation rates.

**Example 2.1**
A scaffold is designed with a size of 4 mm × 4 mm × 4 mm. The scaffold consists of four layers and each layer has two strands, with a cross-view shown below.

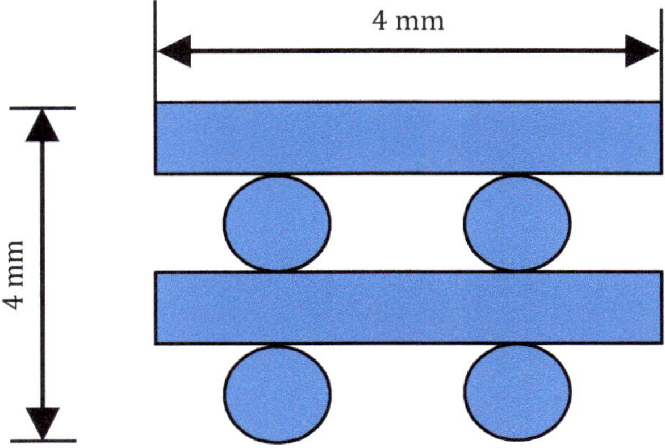

1. Find the strand diameter and porosity of this scaffold.
2. If the design is changed to a scaffold with eight layers, find the strand diameter and number of strands in each layer that are required to achieve the same porosity as obtained in the solution to (1).
3. Find the surface area of all strands in the above two designs, respectively.

**Answer**

1. Ideally, the strand diameter is equal to the layer height. Because the scaffold has a total height of 4 mm and four layers, the strand diameter is 1 mm. The porosity of the scaffold according to Eq. 2.2 would be:

$$
\%\text{Porosity} = \frac{V_{\text{total}} - \sum \text{Volume of all strands}}{V_{\text{total}}} \times 100\%
$$

$$
= \frac{(4 \times 4 \times 4) - 8 \times \left[ \left( \pi \times (0.5^2) \right) \times 4 \right]}{4 \times 4 \times 4} \times 100\% = 60.75\%.
$$

2. The eight-layer scaffold would ideally have strands with a diameter of 0.5 mm. As such, the porosity can be calculated as:

$$
\frac{(4 \times 4 \times 4) - N \times \left[ \left( \pi \times (0.25^2) \right) \times 4 \right]}{4 \times 4 \times 4} \times 100\% = 60.75\%.
$$

Solving the above equation gives $N = 32$. Therefore, the same porosity can be achieved by having 32 strands with a diameter of 0.5 mm. Thus, the number of strands in each layer $= \frac{32}{8} = 4$.

3. The surface area of all strands is $A$, where

$$
A = N \times (\text{Cylindrical surface area} + \text{End surface area}) = N
$$

$$
\times \left( \pi D \times L + \frac{\pi D^2}{4} \times 2 \right)
$$

For the scaffold design in (2):

$$
A = 8 \times \left( \pi \times 1 \times 4 + \frac{\pi \times 1^2}{4} \times 2 \right) = 113.01 \text{ mm}^2
$$

For the scaffold design in (3):

$$
A = 32 \times \left( \pi \times 0.5 \times 4 + \frac{\pi \times 0.5^2}{4} \times 2 \right) = 213.63 \text{ mm}^2
$$

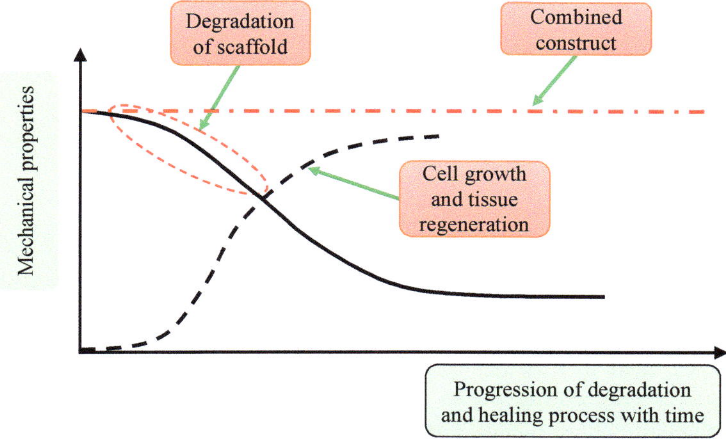


**Fig. 2.3** Relationship between mechanical properties of a scaffold and progression of degradation with time

## 2.2.2  Mechanical Properties

It is generally accepted that the mechanical properties of a scaffold should be similar to those of the tissue/organ to be repaired. Notably, the mechanical properties of scaffolds are not constant but change with time during the tissue repair process. On the one hand, scaffold biomaterials degrade with time resulting in a decrease in mechanical properties. On the other hand, cell growth and tissue regeneration take place and thus impart mechanical properties to the combined construct of the scaffold and cells/tissues. Figure 2.3 shows the progression of degradation and its relationship to mechanical properties. Biomaterials used for scaffold fabrication undergo molecular weight loss and/or degradation. Concurrently, cells grow with the newly regenerated tissues to fill the space created by the degradation of scaffold biomaterials. As such, the mechanical properties are dynamic, depending on the scaffold biomaterial degradation as well as the tissue regeneration. Ideally, the combined mechanical properties should match those of the tissue/organ being repaired during the healing process.

If a mechanical force applied to a scaffold is below the yield strength, the scaffold undergoes elastic deformation. This means it will return to its initial dimensions when the force is removed. If the applied stresses surpass the yield force, the scaffold undergoes plastic deformation and the initially designed structure cannot be maintained. The widely used parameters to characterize the mechanical properties of scaffolds include the elastic modulus and yield strength (as discussed in Chap. 4). Note that the scaffold mechanical properties depend on both the scaffold design and the biomaterial used to fabricate the scaffold.

As noted previously, scaffolds degrade upon implantation, which, consequently, changes their mechanical characteristics. The degradation rate is the rate of scaffold biomaterial degradation under physiological conditions and can be very dissimilar in varying environments (e.g., in vivo vs. in vitro). To examine the degradation rate in vitro, a scaffold is typically weighed at set time points after immersion in a fluid resembling physiological conditions and then evaluated in terms of

$$\%\text{Degradation} = \frac{w_0 - w_t}{w_0} \times 100, \tag{2.3}$$

where $w_0$ is the initial weight of the freeze-dried scaffold and $w_t$ is the weight of the freeze-dried scaffold at a given time point.

Two types of biomaterial degradation mechanisms are considered in printed scaffolds, i.e., bulk degradation and surface degradation, as illustrated in Fig. 2.4. In bulk degradation, scission of the biomaterial (or polymer) chains occurs and progresses by propagating cracks in the scaffold structure. Molecular weight loss also occurs and the mechanical strength of the scaffold decreases with time. In this type of degradation, the decrease in molecular weight occurs as soon as the degradation process begins, while the strand dimensions and pore sizes within the scaffold remain unchanged. In surface degradation, the strand surface degrades and, as a result, the strand dimension and mass decrease with time and also the mechanical strength of the scaffold decreases with time. However, the molecular weight of the

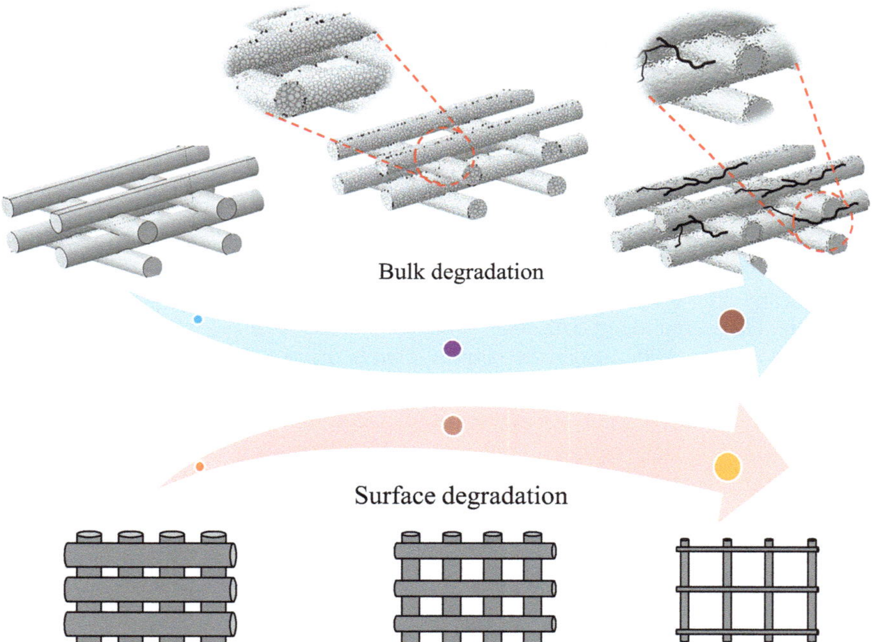


**Fig. 2.4** Bulk and surface degradation of scaffolds over time (arrows show the direction of the degradation process with time)

biomaterial remains unchanged. Surface and bulk degradation often occur simultaneously during scaffold degradation.

Another mechanical property of scaffolds is swelling, which is a measure of the penetration of water into the scaffold structure. The swelling rate of a scaffold can be calculated as follows:

$$\%\text{Swelling} = \frac{w_t - w_0}{w_0} \times 100, \tag{2.4}$$

where $w_t$ is the weight measured at a given time point and $w_0$ is the initial weight of the scaffold. Swelling and degradation are interconnected factors that occur concurrently. Scaffolds begin to swell upon implantation into physiological conditions, which is also the point at which degradation begins to occur.

### 2.2.3  Biological Properties

The biological properties of a scaffold refer to its ability to support cell growth or functions and tissue regeneration, with no or limited negative effects on the host system (i.e., animal or human). The biological properties of a scaffold are typically evaluated by performing in vitro and/or in vivo tests. In vitro (literally "in glass") tests take place in a well-controlled laboratory environment, where scaffolds with living cells are cultured in vials or bioreactors containing simulated body fluid or other appropriate medium. In vivo tests are performed in the living body of an animal or human, with scaffolds implanted at the injured site to help healing. For both in vitro and in vivo tests, samples of scaffolds with newly generated tissue are taken at set time points during the course of a study and then analyzed to evaluate scaffold performance. This includes the examination of cell attachment, survival, proliferation (increasing the number of cells), and differentiation (changing cell types), as well as the evaluation of tissue regeneration, inflammatory response, and by-products of the degraded scaffold. In vitro tests are often performed first and, if the scaffold performance is acceptable, further in vivo tests are then undertaken.

## 2.3  Scaffold Design Process

For tissue engineering or modeling, scaffolds should be designed as 3D porous structures with appropriate architectural, mechanical, and biological properties. The key is to design and/or choose the scaffold architecture, biomaterials, and living cells to be incorporated within scaffolds to achieve the desired results in terms of outcomes. Typically, scaffold design starts from an understanding and/or knowledge of the composition and organization of tissues/organs to be repaired. With such knowledge, scaffolds are designed with appropriate external geometries and internal structures as well as spatial arrangements of biomaterials/cells to mimic the architectural, mechanical, and biological properties of the tissues/organs to be repaired or modelled.


### 2.3.1 Understanding the Composition and Organization of Tissue/Organs

Medical imaging technologies such as computed tomography (CT) and magnetic resonance imaging (MRI) are commonly used to obtain information about the composition and organization of the tissues or organs to be repaired. Medical imaging technologies, as noninvasive modalities, can provide a 3D representation of the structure of tissues/organs, allowing their function to be evaluated at either the small scale (e.g., cellular level) or large scale (e.g., organism level). The acquisition of medical images is the very first step in the design of customized scaffolds for tissue/organ repair [2]. Specifically, CT imaging uses X-rays to penetrate the tissue of interest. Different components of tissues have different rates of X-ray absorption and so can be distinguished as lighter or darker regions on X-ray images. Denser areas, such as bone, appear as white dots in CT scan images. CT imaging starts with rotating the X-ray source around the tissue/organ, measuring the transmitted beam intensity, and recording data that are subsequently used to create cross-sectional (or 2D) images. Such 2D images, with pixels representing a small volume of the tissue/organ, are then compiled to create a 3D construct of the tissue. The same procedure is used for MRI, where a magnetic field instead of X-ray is used to image soft tissues based on nuclear magnetic resonance, aligning a small fraction of the nuclei in the tissue with a magnetic field. MRI has higher contrast resolution than CT imaging, enabling it to provide better spatial resolution.

After acquiring data using either CT or MRI imaging techniques, the data are processed based on tomographic reconstruction techniques, producing 2D images that can be subjected to further evaluations using 3D anatomical representations. Basically, all images obtained from either CT or MRI imaging are stored as 2D data (image slices) based on a threshold of gray values. Using this technique, hard and soft tissues can be distinguished by converting gray values to a 3D model by means of combining 2D imaging data. This process, known as digital geometry processing, creates a 3D volume of the inside of the scanned area. Finally, 3D models of tissue/ organs are generated using mathematical modeling techniques. Using such models, different views of tissues/organs can be observed and volume rendering can be implemented to create a volumetric representation of different segments. The aforementioned models and images can be used to inform computer-aided design (CAD) of scaffolds, as discussed below.

## 2.3.2  *Designing Scaffolds with Appropriate Architectures*

Based on knowledge of the tissue to be repaired, scaffolds must then be designed with appropriate architectures to physically replicate the anatomy of the target tissue or tissue injury site and, as the initial building blocks, to help restore tissue function. As noted previously, the architectural properties of scaffolds include the external geometry and internal structure. As a general rule, a scaffold's external geometry should reflect the geometry of the tissue or injury site, while the internal structure should mimic the porous structure of the target tissue.

Studies have been oriented toward the creation of various internal structures. To this end, different unit cells can be implemented [3–5]; for example, Fig. 2.5 shows rhombicuboctahedron, rhombic dodecahedron, and diamond unit cells as the initial building blocks of tissue/organs used to generate CAD files of artificial tissue/organs with different architectural properties. The challenge here is determining an appropriate internal structure with respect to the target tissue/organ. Generally speaking, the internal architecture of artificial tissue/organs should be designed with appropriate mechanical and biological properties and be able to be produced by current bioprinting and/or other fabrication techniques.

In addition to internal structure, the external geometry of scaffolds is important because it must be appropriate for implantation into the injured site. Figure 2.6 shows the general procedure of creating a 3D model printed from a patient's medical imaging data, which starts with imaging data acquisition by means of CT/MRI, cross-sectional image processing, converting images to the format appropriate for interfacing with printer software, and finally printing the model layer by layer. First, the medical imaging (CT or MRI) data can be converted to the Digital Imaging and

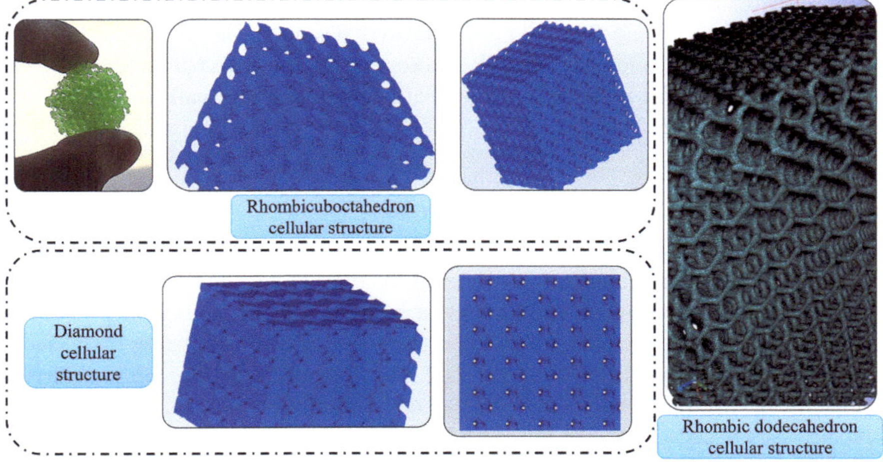


**Fig. 2.5**  Different internal structures of scaffolds generated by a computer: periodic unit cells with rhombicuboctahedron, rhombic dodecahedron, and diamond cellular structures

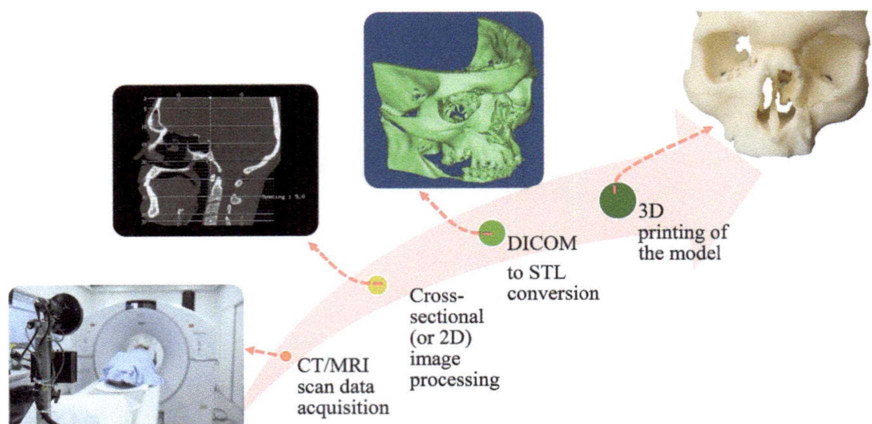



**Fig. 2.6** Procedure for the creation of a customized scaffold: from imaging to printing

Communications in Medicine standard format (DICOM) used in medical digital imaging. A DICOM file includes hundreds of cross-sectional images that must be processed for the next step of DICOM to STL conversion. STL, an abbreviation for stereolithography, is a well-known format for 3D printing and the common format of the CAD models used in bioprinters. An STL file is a combination of hundreds of triangles representing the surface of a model. The process of the creation of a CAD model based on medical imaging is known as segmentation, which can affect the accuracy of the CAD model [6]. In this regard, different software packages use various algorithms, and the accuracy of the CAD model depends on the algorithm implemented to generate the STL file. The STL files are then transferred to bioprinting machines to direct the creation of the layer-by-layer construct. The relationship between the STL file and scaffold design is based on the external geometry of the scaffold. As the STL file represents the external structure of the CAD model, any error in the STL file will affect the resulting scaffold.

### 2.3.3 Selection of Biomaterials/Cells

The selection of biomaterials/cells is critical in scaffold design due to their significant influence on mechanical and biological properties. Cells are building blocks in tissue engineering or modeling with important cellular functions being their viability, proliferation, and differentiation. *Cell viability* refers to the cell's ability to remain alive and functional. Cell death can occur through either necrosis or apoptosis. Necrosis is caused by factors external to the cell or tissue, such as infection or trauma, that result in the unregulated digestion of cell components, changes in chemistry (e.g., decrease in pH), and presence of mechanical forces. In contrast, apoptosis is a naturally occurring programmed and targeted cause of cellular death. While apoptosis often provides beneficial effects to the organism, necrosis is almost always detrimental and can be fatal. *Cell proliferation* is the process by which cells grow in size and/or divide to produce more daughter cells. Cell proliferation leads to an increase in cell size and/or cell number, and thus tissue growth or an increase in size and mass. Typically, cell proliferation involves both cell growth and cell division occurring at the same time. However, sometimes cell growth occurs without cell division, or cell division occurs without cell growth. *Cell differentiation* is the process by which cells change from one type to differentiated ones. The cells with the function of differentiation are termed *stem cells*. Stem cells can differentiate into more specialized ones, such as muscle cells, fat cells, bone cells, blood cells, nervous cells, epithelial cells, immune cells, and sex cells. While differentiating, stem cells undergo a set of changes, usually involving alterations in gene expression, protein synthesis, and phenotype. A cell's *phenotype* refers to its observable characteristics, including its morphology or shape, as well as its ability to produce specific proteins.

Recent advancements in cell culture and biology have opened up the possibility of using different cell sources, including cell lines, primary cells (either from human or animal sources), and progenitor or stem cells, as well as stem-cell derived cells. Many of these cell sources are now commercially available, which allows for the selection of a wide range of cell types in scaffold design. *Cell lines* are transformed cell populations that can divide for an indefinite period due to either immortalization in the laboratory or derivation from a tumorigenic source. Cell lines are quite often used for imperative discoveries due to ease of culture; this is why they are initially used to prove the potential of newly bioprinted structures. Many cell lines are derived from tumorigenic sources, and therefore other cell sources should be implemented for subsequent evaluations to verify the functionality of the bioprinted tissue/organ. *Primary cells*, isolated from living tissue/organs, have a high level of function. Different from stem cells, primary cells are terminally differentiated cells and are a suitable choice for tissue/organ generation due to their functionality. However, primary cells can be relatively hard to culture and have a limited passage number, i.e., the proliferation of these cells gradually declines. Hence, mimicking the in vivo environment is important to achieve efficient functionality as well as maximize the viability of these types of cells. *Stem cells* have the ability to self-renew and differentiate into other types of functional cells. Stem cells are regularly categorized as embryonic stem cells (derived from the inner cell mass of an embryo), adult somatic stem cells (differentiated into the available cell types, residing in that tissue solely), fetal stem cells (derived from fetal environments), and induced pluripotent stem cells (created artificially through reprogramming terminally differentiated cells). All of these stem cells have advantages and drawbacks in bioprinting for tissue engineering. For example, embryonic stem cells can differentiate into any other cell but are limited in their clinical use due to ethical concerns. In addition, the adult stem cells used clinically are partially differentiated and have limited proliferative capacity in vitro. Generally speaking, stem cells have greater potential to mimic the functions of primary cells of a target tissue than cell lines. However, completely differentiating stem cells into the desired primary cells remains challenging.

**Case Study 2.1**

*Cell sources for cardiomyocyte induction and regeneration in myocardial infarction treatment.* Myocardial infarction (MI), commonly known as heart attack, results from interruption of the blood supply to a part of the heart; this is most commonly due to occlusion of the coronary artery. If left untreated, the ischemia (resulting restriction in blood supply) can cause damage or death of the myocardium (cardiac muscle tissue), thus leading to heart failure. Notably, the myocardium has extremely limited regeneration potential due to the loss of or damage to cardiomyocytes (the cells of which the myocardium is comprised), which have an extremely low rate of proliferation that is far from sufficient to compensate for their loss. This is also accompanied by a cardiac remodeling process due to the formation of scar tissue and degradation of the extracellular matrix, eventually leading to heart failure.

With advances in tissue engineering, implantation of bioengineered or bioprinted cardiac patches made from biomaterials into the affected myocardium (Fig. 2.7) shows promise for improving MI repair [7, 8]. This therapy is still being conducted in trials, with one key issue being cardiomyocyte regeneration. In addition to isolated cardiomyocytes, a variety of stem cells and stem cell-derived cardiovascular cells have been exploited to provide renewable cellular sources [7, 9]. Pluripotent stem cells (PSCs), including embryonic stem cells (ESCs) and induced pluripotent stem cells (iPSCs), as well as multipotent mesenchymal stem cells (MSCs) possess trilineage cardiovascular differentiation potential. PSCs are differentiated into cardiac progenitor cells (CPCs) or functional cardiovascular cells by embryonic bodies (EBs) or monolayer cultures and infused after sorting against specific markers to a

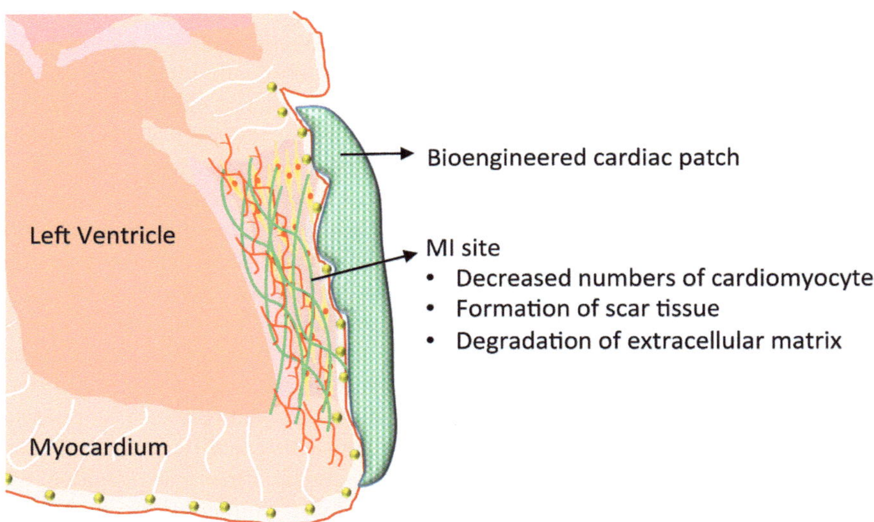


**Fig. 2.7**  Heart with MI treated with an engineered cardiac patch

purity degree of over 95%. Without the risk of teratoma formation and less risk of immune rejection, MSCs can be infused either directly or after in vitro differentiation and sorting. Pre-treatment and infusion with prosurvival factors help to improve the viability of cells upon engraftment. In addition to exogenous cell transplantation, reprogramming of cardiac fibroblasts to functional cardiomyocytes provides an appealing potential method for MI treatment by facilitating the integration of the newly formed cardiomyocytes with surrounding former ones and reducing the formation of fibrotic scar tissue.

Biomaterials include a wide range of polymers, metals, or ceramics. Among these biomaterials, polymers have been extensively used in extrusion-based bioprinting [10, 11]. Polymers are organic biomaterials possessing long chains with high water content and thus are able to provide a hydrated tissue-like environment that supports cell functions (including cell attachment, proliferation, and differentiation) and tissue regeneration. Polymers are categorized as either natural (derived from natural sources) or synthetic (created in a laboratory). Natural polymers are preferred due to their functionality in mimicking the biological nature of the extracellular matrix, while the molecular weight of synthetic materials can be modulated to create hydrogels with desired properties. Natural and synthetic polymers can also be grouped as soft (e.g., hydrogels such as collagen, hyaluonic acid, alginate) and stiff polymers (e.g., thermoplastic polymers such as polycaprolactone (PCL) and polylactic acid (PLA)). Hydrogels are soft polymers with a cell-friendly environment and are often used to incorporate cells into scaffolds. Thermoplastic polymers are stiff polymers that can be melted at relatively high temperatures and, as such, are incompatible with cells and proteins for bioprinting. However, they have been widely used to improve the mechanical properties of printed structures. One thermoplastic polymer commonly used in printing is PCL, which is a biodegradable polyester-based biomaterial. However, it has no binding sites for cell attachment to facilitate tissue integration. Overall, thermoplastic polymer hydrogels, or a combination thereof, can be used or selected for the design of tissue scaffolds. More details on these hydrogels are discussed in Chap. 3.

## 2.4  Typical Scaffold Structures by Bioprinting

In the context of 3D bioprinting, typical scaffolds feature mono, hybrid, or zonal structures. *Mono-structure scaffolds* have a layered structure with a repeated, yet identical, arrangement or strands printed from the same bioink, as illustrated in Fig. 2.8a. These structures have several advantages, such as using bioink made from a single biomaterial or a combination of biomaterials. Hydrogels are widely used for mono-structure fabrication, forming a tissue-like structure. Cells can be easily incorporated into hydrogels due to their favorable environment. However, mono structures quite often have poor mechanical properties. To address this issue, higher hydrogel concentrations can be used to improve the mechanical stability, but this can lead to insufficient in-growth of the target tissue. An alternative to high

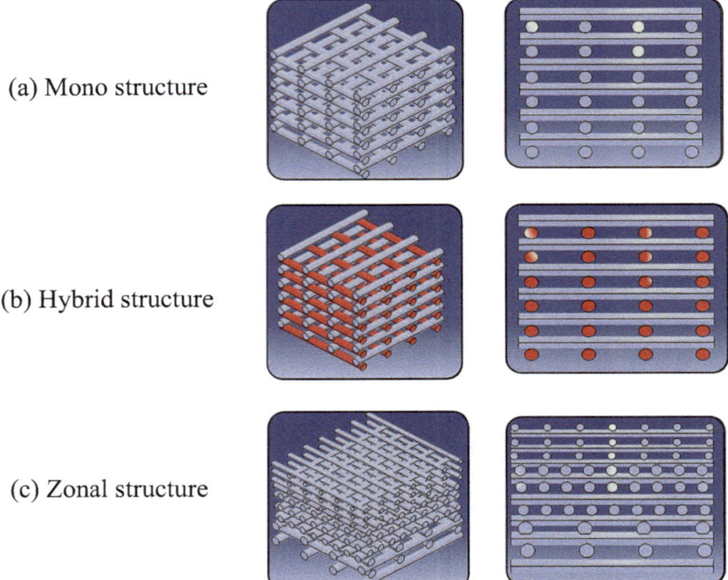



**Fig. 2.8** Perspective (left) and side (right) views of various types of scaffold designs for different applications, including (**a**) mono, (**b**) hybrid (combination of two or more biopolymers in either different or same layers), and (**c**) zonal structures

concentrations is the creation of mono-structure scaffolds from a mix of two or more biomaterials with complementary properties.

*Hybrid-structure scaffolds* are designed and created using two or more biopolymers in either same or different layers (Fig. 2.8b). For instance, a synthetic polymer can be used to improve the mechanical stability while a hydrogel can provide a good environment for cell incorporation. Such a hybrid approach can result in synergies, thus improving the outcomes in both hard and soft tissue engineering applications [12, 13].

*Zonal structure scaffolds* have a zonal structure with design parameters (e.g., porosity in Fig. 2.8c) varying in different zones or in a gradient. Notably, native tissue or organs are typically zonal or gradient structures, broadly classified into cellular, compositional, structural, mechanical, and morphogenic [14]: (1) *cellular gradients* are transitions in the density of one or more cell types, e.g., zone-dependent densities of chondrocytes in articular cartilage; (2) *compositional gradients* involve transitions in the extracellular matrix of tissue, e.g., mineral gradients in tooth dentin; (3) *architectural gradients* are changes in the organization of tissue components, e.g., changes of fiber orientation across cardiac tissue [15]; (4) *mechanical gradients* are variations in mechanical properties (such as compressive, tensile, and shear properties), which are typically found at load-bearing musculoskeletal interfaces such as cartilage; and (5) *morphogenic gradients*, established during the early development stage of a tissue/organ, involve various morphogens or diffusible chemical substances that exert control over morphogenesis including the formation of subsequent cellular, compositional, architectural, and mechanical gradients by spatially regulating cell functions (proliferation, migration, and differentiation). Scaffolds designed with zonal or gradient structures allow for one or more of the above gradient or anisotropic structures of native tissue/organs to be achieved. As such, bioprinting zonal or gradient scaffolds has been drawing considerable attention and shows promise in terms of mimicking native structures and cellular composition and thus enhancing tissue engineering outcomes.

For each of the above structures, the strand diameter/orientation, pore size, and porosity, along with the biomaterials and cells, are key parameters that must be determined during scaffold design. These parameters can affect both the mechanical and biological properties of the scaffold. Scaffold designs based on either a mono, hybrid, or zonal structure are able to provide channels for the creation of vascular networks within their internal structures to enhance the viability of cells. Such channels can provide oxygen and nutrients to cells. More information on the vascularization of scaffolds is discussed in Chap. 7.

**Case Study 2.2**

*Biomimetic design of scaffolds for cartilage tissue engineering.* Osteoarthritis (OA) is a painful, debilitating, and degenerative joint disease characterized by progressive erosion of the articular cartilage that covers the opposing surfaces of bones in diarthrodial joints. The management of OA is difficult because articular cartilage has an extremely limited ability for self-repair. The conventional approach for end-stage OA treatment involves joint replacement with prosthetic implants of artificial material, but suffers from inferior functionality compared to native cartilage as well as the need for additional surgeries (or revisions) to address complications such as misalignment or degradation of the artificial material. Cartilage tissue engineering aims to create cartilage tissue substitutes that closely mimic the biomechanical properties of native articular cartilage and provide a milieu that augments endogenous tissue repair mechanisms for the repair of injured or diseased joint cartilage [16, 17].

Notably, native articular cartilage has a zonal or graduate structure and cellular composition. As articular cartilage transitions from the superficial zone to middle, deep, and calcified zones (Fig. 2.9a), the extracellular matrix of the cartilage is characterized by distinct orientations of collagen fibers (changes from parallel to the surface or the relative movement direction at a joint to perpendicular to the surface), decreased cell (chondrocyte) density, as well as increased glycosaminoglycans/ mechanical strength/oxygen tension/nutrient availability [16–18]. Such a unique gradient and zonal structure enables and/or supports the healthy function of articular cartilage, which features low-friction lubrication at the superficial zone, excellent shock absorption in the middle zone, and great permeability and strong mechanical support in the deep/calcified zones.

Bioprinting biomimetic constructs has been an emerging and promising approach to mimic native articular cartilage. Among these constructs are those with biomimetic structures that feature zone-dependent architecture, mechanical properties, and

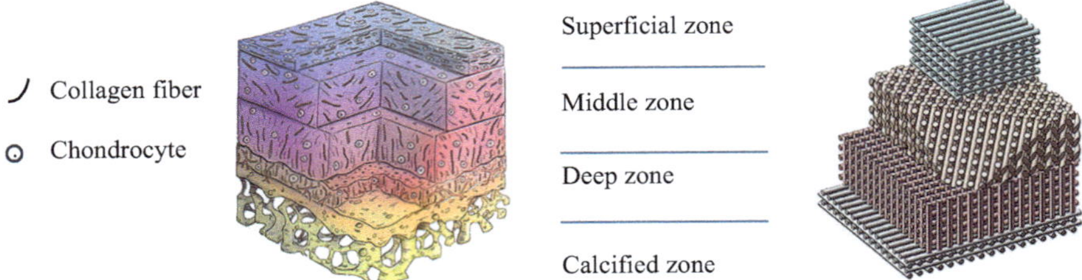


**Fig. 2.9** (**a**) Zonal structure of articular cartilage, showing the transitions from the superficial zone to middle, deep, and calcified zones; (**b**) a scaffold design to mimic the zonal structure of native cartilage

cellular composition. As an example, Fig. 2.9b shows the design of a scaffold to mimic the zonal structure of native cartilage. The scaffold is comprised of four zones, i.e., superficial, middle, deep, and calcified zones, with unique orientations of printed strands to mimic the structure of native cartilage. Specifically, the strands run parallel to the articular surface in both the superficial and calcified zones, run diagonally in the middle zone, and run vertically in the deep zone. Though promising, such biomimetic constructs for cartilage tissue engineering are still being evaluated in trials with many issues to be addressed.

## 2.5  Summary

The important requirements for tissue engineering scaffolds can be classified into three categories, i.e., architectural, mechanical, and biological. From the architectural point of view, scaffolds feature both an external geometry and internal structure. A scaffold's external geometry should be appropriate for implantation into the injured site, while the internal structure should be highly porous with interconnected pores for facilitating nutrient transport and metabolic waste removal during the repair process. Pore size and porosity are commonly used parameters to characterize scaffold architecture.

With respect to mechanical properties, scaffolds should have sufficient mechanical stability upon implantation to the injured site. Notably, the mechanical strength of scaffolds during the repair process is not constant but changes with time. On the one hand, scaffold biomaterials degrade with time and, as a result, the mechanical strength of the scaffold will decrease. On the other hand, cell growth and tissue regeneration take place and thus impart mechanical strength to the combined construct of scaffold and cells or tissues.

The biological properties of a scaffold refer to its ability to support cell growth or function and tissue regeneration, with limited or no negative effects on the host system (i.e., animal or human), which is evaluated through in vitro and/or in vivo tests.

Generally speaking, scaffolds should be designed with 3D porous structures that will impart appropriate mechanical and biological properties for the intended application. The scaffold design process mainly involves designing and/or choosing the scaffold architecture, biomaterials, and living cells to be incorporated. Typically, scaffold design starts from an understanding and/or knowledge of the composition and organization of the tissues/organs to be repaired. With such knowledge, scaffolds can be designed with appropriate external geometries and internal structures and also biomaterials/cells selected and spatially arranged so as to mimic the architectural, mechanical, and biological properties of the tissues/organs to be repaired. Knowledge with respect to the composition and organization of tissues or organs to be repaired is commonly obtained using medical imaging technologies such as CT and MRI. Based on knowledge of the tissue to be repaired, scaffolds must be designed with appropriate architectures to physically replicate the anatomy of the target tissue and serve as the initial building block to restore tissue function. In this regard, the patient's medical imaging data can be used to design the external geometry while various types of units can be used as the internal cellular structure. Once the internal and external architecture of a scaffold are defined, biomaterials and cells are selected, while keeping in mind their effects on the mechanical and biological properties.

Natural or synthetic polymers (including soft and stiff polymers such as hydrogels and thermoplastic polymers, respectively) can be selected as the main scaffold matrix, to which cells are then incorporated. Cells are building blocks in scaffolds, where their cellular functions of viability, proliferation, and differentiation play important roles. Cell sources include cell lines, primary cells, and stem cells.

In the context of 3D bioprinting, typical scaffolds have either mono, hybrid, or zonal structures. Mono-structure scaffolds feature a layered structure with a repeated, yet identical, arrangement of strands printed from the same bioink. Such structures are often made of hydrogels and have poor mechanical properties despite their cell-friendly environment. Hybrid structures are created using a combination of two or more biopolymers in either similar or different layers. A synthetic polymer generally improves the mechanical stability of a hybrid scaffold while another component, such as a hydrogel, can provide a good environment for cell incorporation and a highly hydrated and tissue-like network structure. Hybrid structures capitalize on the synergies between two (or more) components to achieve the desired results. Zonal or gradient structures feature various design parameters (e.g., porosity) and biomaterials/cells in different zones or in gradients. Such zonal or gradient scaffolds allow for better mimicking the gradient or anisotropic structures of native tissue or organs, which are broadly classified into cellular, compositional, structural, mechanical, and morphogenic gradients.

**Problems**
1. Briefly describe the general requirements imposed on scaffold design from the architectural, mechanical, and biological perspectives.
2. Briefly explain how the mechanical properties of scaffolds, once implanted, change with time, along with the tissue regeneration within the scaffolds.

3. Briefly explain the scaffold design process.
4. Briefly explain the cell functions important to tissue engineering, as well as the common cell types for use in tissue scaffolds.
5. Briefly explain three structures typically employed in scaffold designs by means of bioprinting, along with their pros and cons.
6. From the literature, identify a case study where the scaffolds were designed to mimic the structure and cellular composition of native tissue or organ.
7. Calculate the porosity of the scaffold shown in the following figure. The strand diameter is 0.5 mm and other dimensions are shown in the figure.

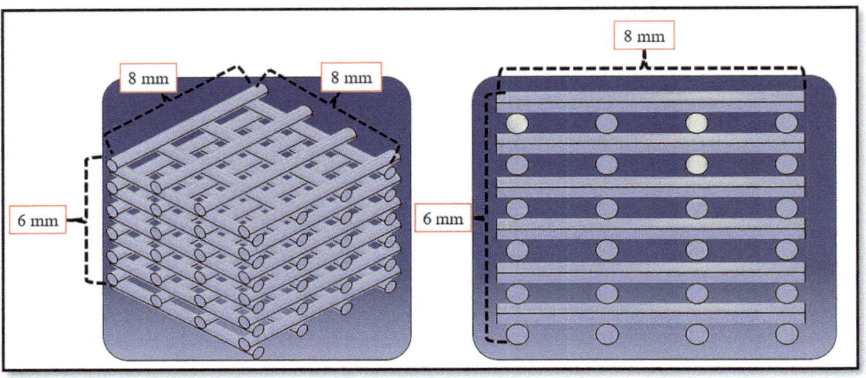


# References

1. Z. Li, Z. Chen, X. Chen, R. Zhao, Multi-objective optimization for designing porous scaffolds with controllable mechanics and permeability: A case study on triply periodic minimal surface scaffolds. Compos. Struct. **333**, 117923 (2024). https://doi.org/10.1016/j.compstruct.2024.117923
2. L. Ning, X. Chen, A brief review of extrusion-based tissue scaffold bio-printing. Biotechnol. J. **12**(8), 1600671 (2017). https://doi.org/10.1002/biot.201600671
3. X. Chen, Dispensed-based bio-manufacturing scaffolds for tissue engineering applications. Int. J. Eng. Appl. **2**(1), 10–19 (2014)
4. R. Hedayati, M. Sadighi, M. Mohammadi-Aghdam, A.A. Zadpoor, Mechanical properties of regular porous biomaterials made from truncated cube repeating unit cells: Analytical solutions and computational models. Mater. Sci. Eng. C **60**, 163–183 (2016). https://doi.org/10.1016/j.msec.2015.11.001
5. Z. Li, Z. Chen, X. Chen, R. Zhao, Design and evaluation of TPMS-inspired 3D-printed scaffolds for bone tissue engineering: Enabling tailored mechanical and mass transport properties. Compos. Struct. **327**, 117638 (2024). https://doi.org/10.1016/j.compstruct.2023.117638
6. S. Naghieh, A. Reihany, A. Haghighat, et al., Fused deposition modeling and fabrication of a three-dimensional model in maxillofacial reconstruction. Regen. Reconstr. Restor. **1**, 139–144 (2016). https://doi.org/10.22037/rrr.v1i3.12543

7. L. He, X. Chen, Cardiomyocyte induction and regeneration for myocardial infarction treatment: Cell sources and administration strategies. Adv. Healthc. Mater. **9**, 2001175 (2020). https://doi.org/10.1002/adhm.202001175

8. M. Izadifar, D. Chapman, P. Babyn, et al., UV-assisted 3D bioprinting of nanoreinforced hybrid cardiac patch for myocardial tissue engineering. Tissue Eng. Part C Methods **24**, 74–88 (2018). https://doi.org/10.1089/ten.tec.2017.0346

9. F. Ketabat, J. Alcorn, M.E. Kelly, et al., Cardiac tissue engineering: A journey from scaffold fabrication to in vitro characterization. Small Sci. (2024). https://doi.org/10.1002/smsc.202400079

10. S. Naghieh, M. Sarker, M. Izadifar, X. Chen, Dispensing-based bioprinting of mechanically-functional hybrid scaffolds with vessel-like channels for tissue engineering applications – A brief review. J. Mech. Behav. Biomed. Mater. **78**, 298–314 (2018). https://doi.org/10.1016/j.jmbbm.2017.11.037

11. X.B. Chen, A. Fazel Anvari-Yazdi, X. Duan, et al., Biomaterials/bioinks and extrusion bioprinting. Bioact. Mater. **28**, 511–536 (2023). https://doi.org/10.1016/j.bioactmat.2023.06.006

12. Z. Izadifar, T. Chang, W. Kulyk, et al., Analyzing biological performance of 3D-printed, cell-impregnated hybrid constructs for cartilage tissue engineering. Tissue Eng. Part C Methods **22**, 173–188 (2016). https://doi.org/10.1089/ten.tec.2015.0307

13. C.H. Park, H.F. Rios, Q. Jin, et al., Biomimetic hybrid scaffolds for engineering human tooth-ligament interfaces. Biomaterials **31**, 5945–5952 (2010). https://doi.org/10.1016/j.biomaterials.2010.04.027

14. C. Li, L. Ouyang, J.P.K. Armstrong, M.M. Stevens, Advances in the fabrication of biomaterials for gradient tissue engineering. Trends Biotechnol. **39**, 150–164 (2021). https://doi.org/10.1016/j.tibtech.2020.06.005

15. F. Ketabat, T. Maris, X. Duan, et al., Optimization of 3D printing and in vitro characterization of alginate/gelatin lattice and angular scaffolds for potential cardiac tissue engineering. Front. Bioeng. Biotechnol. **11**, 1161804 (2023). https://doi.org/10.3389/fbioe.2023.1161804

16. Z. Izadifar, X. Chen, W. Kulyk, Strategic design and fabrication of engineered scaffolds for articular cartilage repair. J. Funct. Biomater. **3**, 799–838 (2012). https://doi.org/10.3390/jfb3040799

17. F. You, B.F. Eames, X. Chen, Application of extrusion-based hydrogel bioprinting for cartilage tissue engineering. Int. J. Mol. Sci. **18**, 1597 (2017). https://doi.org/10.3390/ijms18071597

18. C.J. Little, N.K. Bawolin, X. Chen, Mechanical properties of natural cartilage and tissue-engineered constructs. Tissue Eng. Part B Rev. **17**, 213–227 (2011). https://doi.org/10.1089/ten.teb.2010.0572

# Chapter 3
# Biomaterials and Bioinks for Bioprinting

## 3.1 Introduction

Bioprinting is a process of printing or applying bioink in a predesigned manner to build up 3D constructs in a layer-by-layer fashion. A bioink is the formulation of biomaterials, living cells, and/or bioactive molecules. Biomaterials can be classified into three categories, that is, metal, ceramics, and polymers. Among these materials, polymers have been commonly used in bioprinting because they are able to form hydrated and structurally stable 3D environments similar to natural extracellular, allowing for cell functions and growth. Based on the source of polymeric biomaterials that comprise the backbone, hydrogels are classified as either natural or synthetic. Polymer solutions can be crosslinked or gelled under suitable conditions (e.g., UV light, ionic, pH, and temperature) to form a structurally stable polymeric network or hydrogel.

During the extrusion bioprinting process, the bioink is loaded into the bioprinter cartridge and then extruded to form 3D constructs with the help of suitable crosslinking reactions or mechanisms. Successful bioink bioprinting of constructs and their subsequent applications rely on the properties of the formulated bioinks, including printability, crosslinking capability, biological properties, and mechanical properties. Printability and crosslinking capability directly affect if the formulated bioinks can be printed to form 3D structures, while biological and mechanical properties are closely associated with subsequent applications of the bioprinted constructs. This chapter presents the biomaterial properties that are important for bioprinting, followed by a discussion of the natural, synthetic, and composite polymers or hydrogels commonly used in the extrusion bioprinting process.

© The Author(s), under exclusive license to Springer Nature Switzerland AG 2025

D. X. B. Chen, *Extrusion Bioprinting of Scaffolds for Tissue Engineering*,

https://doi.org/10.1007/978-3-031-72471-8_3

## 3.2   Important Properties of Biomaterials for Bioprinting

### *3.2.1   Printability*

In the context of 3D extrusion printing, the printability of a bioink or a biomaterial solution refers to its ability to form and maintain a 3D structure by printing. During the bioprinting process, the solution is printed on a supporting or printing stage and then stabilized by crosslinking or gelation to form a 3D structure. During the printing and before gelation, the bioink is in a solution or semi-solution form and is able to flow or spread on the surface of the printing stage. As a result, the printed structure becomes different from that designed; this structural difference is measured and characterized by the term of printability, as shown in Fig. 3.1. In some cases, the printed structures may collapse and fail to form a 3D structure; such cases are deemed unprintable.

Standardized methods to quantify printability have yet to be defined, the common practice is to examine the printability in terms of extrudability, filament fidelity, and structural integrity [1–5]. Extrudability refers to the capability to extrude bioink through a needle to form a continuous and controllable filament, with more details discussed in Sect. 6.3.5. Filament fidelity refers to the difference in cross-section between the filament formed on the printing stage and the needle for printing (typically, a circle one). Due to the spreading or flow of the bioink, once printed, the cross-sectional profile of the filament formed varies depending on the layer and location, as shown in Fig. 3.1. Structural integrity defines the capability to maintain the 3D structure post printing. Structural integrity is degraded as the printed filaments become fused at their intersection or deformed hanging-over the adjacent

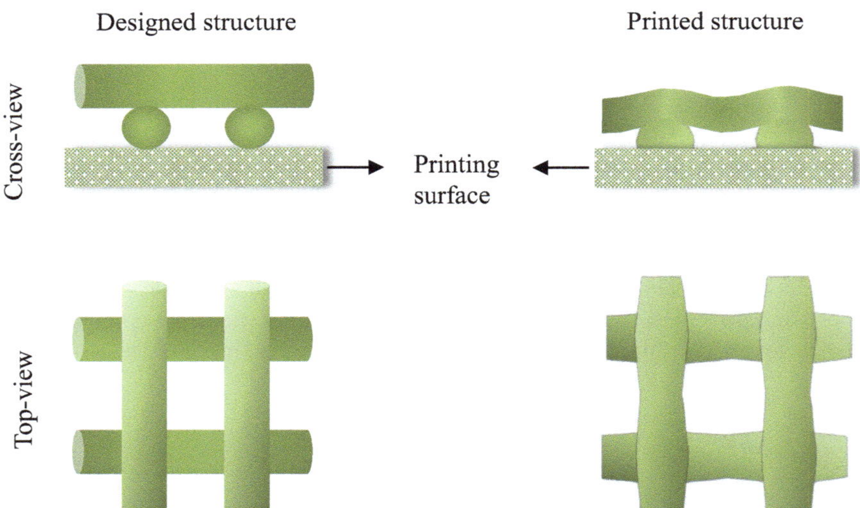


**Fig. 3.1**  Difference between designed and printed structures

layers below, which results in the changes (from the designed structure) in the thickness of each individual layer, height of the whole construct, and/or pore size in all directions, as seen in Fig. 3.1. The printability of a bioink can be affected by many factors, which are generally associated with the formation of the first layer on the printing stage and its physical properties (such as contact angle), flow behavior, and crosslinking mechanisms.

### 3.2.1.1  First Layer Formation

The first layer of printed bioink plays an important role in the printing process. Figure 3.2 shows two strands formed with different contact angles, or the angle between the printed strand profile and the printing surface. A large contact angle helps maintain the fidelity of the printed strand or similarity to the printed cylinder shape, while a small contact angle helps anchor the printed structure on the printing stage and avoid undesired moves and possible deformation during the printing process, thus maintaining the integrity of the printed structure. It is desired for the print stage to have high surface roughness so that the first layer of the printed construct can stick to the surface to maintain the overall structural integrity and stability. Notably, most of the surface of the printing stage, such as glass or plastic ones, have large contact angles with the printed hydrogel and, as such, establishing the anchor between the printed structure and printing surface is difficult. This issue can be overcome by either printing hydrogels in a hydrophobic high-density media, such as perfluorotributylamine, to decrease the contact angle when printing, or coating the printing surface with a thin layer of chemicals [6–8], such as 3-(trimethoxysilyl) propyl methacrylate or polyethyleneimine, to modify the printing surface properties for decreased contact angle.

### 3.2.1.2  Flow Behavior

The flow behavior of a biomaterial solution is associated with its resistance to flow and is characterized by the relationship between the shear stress and shear rate within the solution. Typically, the flow behavior is measured in terms of viscosity, which is

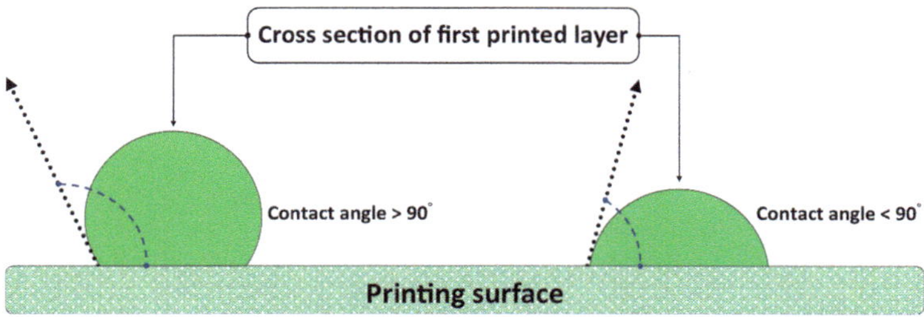


**Fig. 3.2** First layer formed in the bioprinting process

the ratio of shear stress to shear rate. More details regarding the characterization of the flow behavior of biomaterial solutions are provided in Chap. 5.

The flow behavior of a bioink has a significant effect on its printability. The more viscous the bioink solution, the better printability it has. This is due to the fact that a more viscous bioink, once printed, is more difficult to flow or spread, thus being easier to maintain its printed cylinder shape. However, encapsulated cells may survive better in less viscous bioink solutions. In addition, more viscous bioinks require higher pressures for bioprinting. As a result, cells in the bioink are exposed to high process-induced forces, such as shear forces, which negatively influence cell survival and functions [9, 10], as discussed in Chap. 6.

### 3.2.2  Crosslinking Mechanisms

An important procedure in the bioprinting process is the transition from a polymer solution to a gelled or crosslinked hydrogel. By the crosslinking process, the polymer chains are joined together by either physical methods and/or chemical reactions, thus producing hydrogel with a structurally stable polymeric network. The most commonly used crosslinking methods for bioprinting are ionic, thermal, and photo crosslinking, as listed in Table 3.1.

*Ionic crosslinking* is a physical one that occurs when a water-soluble and charged polymer crosslinks with ions of opposite charge. Alginate, for example, is a well-known example of a polymer that can be crosslinked by divalent metal ions, such as $Ca^{2+}$, $Ba^{2+}$, and $Zn^{2+}$ [11]. Ionic crosslinking is an important mechanism in bioprinting as it provides mild and instant gelation of the polymer solution. Its drawbacks include the limited printabilily and mechanical strength, as well as the release of metal ions into the body after implantation. For ionic crosslinking, water-soluble calcium salts such as calcium chloride ($CaCl_2$), calcium sulphate ($CaSO_4$), and calcium carbonate ($CaCO_3$) are commonly used for crosslinking. Addition of $Ca^{2+}$ ions (or other di/trivalent cations) causes rapid gelation of solution. Because

**Table 3.1**  Common crosslinking methods and their characteristics

|  | Ionic crosslinking | Thermal crosslinking | Photo crosslinking |
| --- | --- | --- | --- |
| Methods | Crosslinked by divalent or tri-valent cations (e.g., $Ca^{2+}$) | Crosslinked below the gel transition | Crosslinked by light or UV light |
| Examples of polymers | Alginate, alginate-Gelatin | Agarose, Gelatin, collagen, MC | Gelatin-methacry-late, collagen, fibrin |
| Characteristics | Rapid gelation<br>Under mild conditions<br>Mechanical weakness<br>Metal ions release | Sensitive to temper-atures<br>Long crosslink time<br>Mechanical weak-ness<br>Cell damage | Strong mechanical properties<br>Cell damage |

this crosslinking happens instantaneously under physiological conditions, ionic-crosslinked hydrogel have been widely employed in bioprinting for tissue engineering applications [11, 12].

*Thermal crosslinking* occurs in polymers that are sensitive to temperature, and increasing or decreasing temperature can lead to crosslinking or gelation. Polymers forming hydrogels through thermal crosslinking, such as agarose, gelatin, and collagen, have a gel transition temperature below which the solution gels. Agarose, for example, has a chain structure of random coil conformation at a temperature higher than 40° but changes to have a helical structure when gelling if temperature drops to 32°. These polymers dissolve when heated, and their solutions can be printed on a cooling stage whereupon the polymer passes through its gel transition point and gels [5, 8]. Typically, gels formed by thermal crosslinking are mechanically weak. When applied as bioinks for bioprinting, important considerations include: (1) the melting temperature and gelation temperature of bioink must be compatible with embedded cells as high temperature may cause cell damage and malfunctions after bioprinting and (2) the melting temperature and gelation temperature of bioink must be appropriate for the temperature control in the bioprinting process. Notably, for a bioink with a small difference between melting temperature and gelation temperature, small temperature drops can significantly increase its viscosity. As such, the high pressure is required to expel the bioinks, which might affect the viability of embedded cells. In the other hand, for a bioink with a large difference between melting temperature and gelation temperature, the time required for the solution to gel increases, resulting in more flow or spread of the printing bioink on the printing stage and thus degrading the printed structures.

*Photo crosslinking* refers to the photo-induced formation of a covalent bond between macromolecules to form a crosslinked network. Photo-curable polymers can be printed to form 3D hydrogels if illuminated by laser or visible light. Some other polymers, such as proteinaceous biopolymers, which contains tyrosine residues (such as collagen, fibrin, and gelatin), can be photo-crosslinked only if an appropriate photoinitiator is incorporated [13]. Although many polymers cannot be directly crosslinked by light, a chemical reaction with an acrylate or methacrylate-based agent makes them photo-crosslinkable. These polymers are usually crosslinked by UV light, typically, in 320–365 nm. Notably, the use of UV light has potential biological risks as the UV light may damage cells in the printed constructs and may also be harmful to the operators. Photo crosslinkable polymer bioprinting has been reported by using gelatin-methacrylate, hyaluronan-methacrylate, and dextran-methacrylate [14].

Printability is influenced by the crosslinking capability of bioink, as well as by the bioprinting process itself. Immediate crosslinking of bioink upon printing is a common method used to maintain the printed structures. Besides, other methods have also been developed for the bioprinting process to control and facilitate the bioink crosslinking, as introduced in Sect. 6.4.3. Notably, printed scaffolds with insufficiently crosslinked hydrogel structures might cause the structure change in the printed constructs and even face structural collapse.

## *3.2.3  Biological Properties and Biodegradation*

### 3.2.3.1  Biocompatibility

Biocompatibility must be considered prior to the use of any material for bioprinting. Biocompatibility is defined as the ability of a material to perform with an appropriate host response in a specific situation. Specifically, the biocompatibility of a material encompasses three aspects: (i) does not change the structure and function of cells and tissues it comes in direct contact with, that is, cytocompatibility; (ii) degradation products should also be cytocompatible; and (iii) when implanted in vivo, the material and its degradation products do not induce any immune rejection response. With the advent of tissue engineering, the definition of biocompatibility has evolved and can now also include active interaction with endogenous tissues and/or the immune system and even the support of appropriate cellular activity. Under this paradigm, the biocompatibility of hydrogels should be considered from the time of bioprinting and in vitro maturation (cytocompatibility) through to implantation (immunogenicity) and long-term effects (degradation byproducts).

First, let's consider the biocompatibility of a hydrogel through the bioprinting process. An critical difference between bioprinting and conventional scaffold fabrication techniques lies in that living cells are incorporated in the fabrication process. Living cells are mixed with the hydrogel-forming polymer solution to form bioink and then go through the bioprinting process. During the process, cells are exposed to the process-induced stresses, such as shear stress, which possibly causing cell damage. As a result, only a part of cells in the bioink can survive the bioprinting process. Cell viability refers to the percentage of survival cells to the total number of cells incorporated, and a widely used technique to determine cell viability is a live/dead assay. The cell viability of extrusion bioprinting varies in a wide range, depending on the cell types [1, 5, 9, 15]. After bioinks have been deposited, hydrogel composition can play a more important role in supporting cell viability and proliferation. Some naturally derived polymers have cell-adhesive peptide sequences, which can provide a conductive microenvironment that improves cell viability and proliferation compared to synthetic polymers. Therefore, many efforts have been made to modify synthetic hydrogels with cell-adhesive sequences to achieve better cell viability, proliferation, and differentiation.

Bioprinted constructs are expected to be implanted in vivo after fabrication. The cells, as well as the hydrogels, are both potential antigenic sources. Two mechanisms are involved in the immune response to a foreign body: (1) innate immunity caused by macrophages, neutrophils, and natural killer cells and (2) acquired immunity caused by T and B lymphocytes. The innate immune system is responsible for a nonspecific foreign body reaction, leading to the infiltration of fibroblasts, endothelial cells, and macrophages that form an ensuing fibrotic capsule to isolate the implanted material. In contrast, the acquired immune system would generate an antigen-specific reaction. Naturally derived biomaterials are susceptible to acquired immunity owing to the presence of antigens, while synthetic biomaterials are usually susceptible to innate immunity. The immunogenicity of a biomaterial is important because an intense immune response will lead to shorter scaffold degradation times, a potential attack on the embedded cells, and a higher possibility of fibrosis other than tissue regeneration.

### 3.2.3.2  Biodegradation

Biodegradation is an important property of tissue engineered scaffolds. Bioprinted hydrogels should degrade to monomers that are water-soluble, nontoxic, and can be metabolized by the liver and/or excreted via the kidney. Moreover, the degradation mechanisms and byproducts obtained should not elicit harmful changes that cause damage to the regenerating tissue and/or surrounding tissues.

Hydrogels formed in bioprinting can be degraded by water-induced cleavage of certain bonds/chains (or hydrolysis) in an aqueous environment and/or by enzymatically induced cleavage with the presence of specific enzymes. Biodegradation through hydrolysis usually occurs in bulk and/or surface degradation mechanisms, as schematically shown in Fig. 3.3. In bulk degradation, the rate of water entering the hydrogel is much larger than the one at which the hydrogel is converted into its (water-soluble) degradation products. As a result, the hydrogel maintains approximately the same gross size, while the cleavage of chains in hydrogel undergoes increasing with time and the resulting mechanical properties of the hydrogel decrease rapidly with time, as illustrated by the stress-strain curve beside. In surface degradation, the rate of water ingress into the hydrogel is much less than the one of hydrogel hydrolysis. Consequently, only degrades the surface area of the hydrogel, leading to the decrease in its size. Meanwhile, the mechanical properties of the hydrogel changed little with time. In enzymatic degradation, the bond/chains cleavage in hydrogel is caused by the catalytic action of enzymes under abiotic conditions, and its rate depends on both the number of cleavage sites of the hydrogel and enzyme concentration. Chitosan, for example, can be degraded by lysozyme, producing nontoxic oligosaccharides that can then be excreted and/or incorporated to glycosaminoglycans and glycoproteins.

Control over the degradation rate of hydrogels can be achieved in many ways. The first one is to regulate the cell/polymer ratio. Cells are the source of matrix remodeling proteases, and thus relatively lower cell densities or higher polymer concentrations can decrease the degradation rates, thus extending degradation times. However, lowering the cell densities can also lead to poor tissue regeneration. Another way to adjust the hydrogel degradation time is by controlling the crosslinking degree of the polymeric network. Increasing polymer concentration, crosslinking agent concentration, and exposure time of the crosslinking agent are methods to achieve higher crosslinking and thus slower degradation rates. Researchers have also modified the hydrogel-forming polymers with peptides that are sensitive to enzymatic degradation to achieve control over the degradation behavior of hydrogels. The degradation behavior of hydrogels is mainly determined by calculating weight changes with respect to their initial condition.

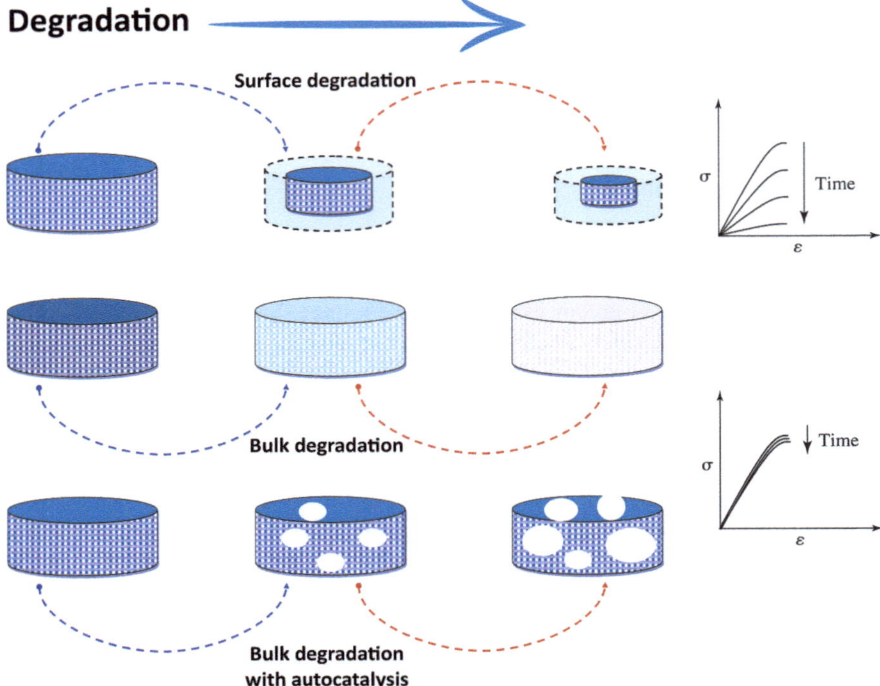



**Fig. 3.3** Hydrogel degradation process, along with the changes in the sample size and stress-strain curve

## 3.2.4 Mechanical Properties

The mechanical properties of a material mainly refer to mechanical strength and are characterized by the relationship between the applied force and the resulting deformation. The mechanical properties of a biomaterial have a significant effect on its printability and thus on the resulting scaffold structure and properties. Sufficient mechanical strength of a biomaterial is essential to maintain the structural integrity of printed scaffolds through the printing process and afterward, so as to provide the mechanical support required for cell growth and tissue regeneration. More details regarding the characterization of mechanical properties of biomaterials are provided in Chap. 4.

## 3.3 Biomaterials for Bioprinting

Hydrogel-forming biomaterials can be divided into two major classes based on their source, that is, naturally derived and synthetic. These materials serve as an important component of bioinks and are used to embed living cells for bioprinting. Most hydrogels used for formulating bioinks are derived from natural extracellular matrix sources, such as collagen, gelatin, fibrin, hyaluronic acid (HA), alginate, and agarose. These naturally derived polymers have intrinsic capabilities to support cell viability and proliferation and can be relatively easy degraded or metabolized. However, the limitations of natural biomaterials include rejection and/or immune-related sequelae, quick degradation rates, and poor mechanical properties. Synthetic hydrogels are usually biologically inert and nonbiodegradable but can be modified to overcome the disadvantages of natural hydrogels. The following section describes the natural or synthetic hydrogels commonly used in bioprinting.

### *3.3.1 Natural Hydrogels*

*Alginate*, or alginic acid, is a water-soluble polysaccharide primarily derived from brown seaweeds. This family of natural polymers is comprised of β-D-mannuronic acid (M) and α-L-guluronic acid (G). The monomers can appear in homopolymeric blocks of consecutive G-residues (G-blocks), consecutive M-residues (M-blocks), or alternating M- and G-residues (MG-blocks). Varying amounts of G and M blocks in alginate result in molecular weights that can range from 50 to 100,000 kDa.

Alginate has been extensively used in extrusion bioprinting due to its water absorbency, low price, and crosslinking rate. Researchers have taken advantage of the rapid crosslinking speed of alginate when it encounters multivalent cations (e.g., $Ca^{2+}$, $Ba^{2+}$) to build 3D scaffolds. These cations interact with the carboxylic groups in alginate and form a gel network. The crosslinking process can be facilitated by spraying the crosslinking agent (e.g., calcium chloride, calcium carbonate, or calcium sulfate) on the printed alginate or printing the alginate into a reservoir containing the crosslinking agent. One drawback of the spraying method is that atomized agents can be trapped by the alginate during bioprinting, resulting in structural deformation and instability of the printed constructs. Therefore, this method is more suitable for alginate solutions with high viscosity. The method of printing alginate into a crosslinking agent reservoir is compatible with both high and low viscosity alginate solutions because alginate polymerization is triggered once it is printed into the agent solution, which provides adequate cations to facilitate the gelation process and thus ensures the stability of the alginate structure.

Alginate solutions also can be weakly pre-crosslinked before loading into the syringe for bioprinting. This method aims to improve the viscosity of the alginate in the solution preparation and also has the potential to protect cell viability during the bioprinting process. In pre-crosslinking, crosslinking agents with a low concentration of cations are mixed into the alginate, leading to slight alginate gelation that increases the solution viscosity. Pre-crosslinking increases the deposition quality of alginate and promotes the integrity of the scaffold after bioprinting, when the scaffold is exposed to a higher concentration of the crosslinking solution. However, a higher pressure is required to extrude or dispense the pre-crosslinked material due to the increased viscosity, and achieving uniform pre-crosslinked alginate solutions can be challenging.

Alginate solutions used in scaffold bioprinting have demonstrated compatibility with many cell types from different tissues, such as bone, muscle, cartilage, skin, nerve, and blood vessels as well as functional organs including the liver, kidney, and bladder. However, alginate possesses the critical disadvantage of poor cell adhesion. Lack of adhesion molecules in alginate, or transmembrane glycoproteins, significantly reduces the interaction between cells and alginate and thus cell functions. The adhesion properties of alginate can be improved by adding other biomaterials that have inherent capacities for cell attachment during alginate solution preparation or modifying the alginate solution by special adhesion molecule sequences, such as RGD peptides, that can covalently bond to alginate chains.

*Chitosan* is a deacetylated derivative of chitin, a common natural polymer found in crustacean shells and fungi cell walls. It is a linear polysaccharide composed of randomly distributed N-acetyl-D-glucosamine (acetylated unit) and $\beta$-(1–4)-linked D-glucosamine (deacetylated unit) that is well known for its nontoxic, biodegradable, and antibacterial properties; as such, it has been used in many medical applications ranging from drug delivery to wound dressings [16], as well in bioprinting [17, 18].

Chitosan is not soluble in neutral pH water but can be dissolved in acid solutions (e.g., acetic acid). Ionic and covalent crosslinking of chitosan solutions both result in the formation of hydrogels. For example, chitosan hydrogels can be readily formed due to ionic interactions between cationic chitosan and negatively charged molecules such as sulfates, citrates, and phosphates ions; however, typical problems that occur with ionic crosslinking are limited mechanical properties and inconsistent performance. The other method to form permanent chitosan hydrogel networks is covalently bonding the polymer chains using crosslinking agents such as $-NH_2$ and $-OH$ groups, which form a number of linkages between chitosan chains including amide and ester bonds.

As an acid environment is not suitable for cell survival, bioprinting cell-incorporated scaffolds using typical chitosan solutions and living cells is challenging. One way to solve the problem is chemically modifying the properties of chitosan to make it soluble in water, with neutral pH after dissolution. The utilization of chitosan is also often limited by its slow gelation rate and poor mechanical properties for bioprinting. These limitations can be alleviated by adding other hydrogels to chitosan solutions to enhance the polymerization rate and structural strength.

*Agarose* is a water-soluble polysaccharide that is purified from seaweed. It is a thermosensitive material that can self-crosslink and de-crosslink using temperature control. Based on various degrees of hydroxymethylation, which determine the temperature range for the crosslinking and de-crosslinking of agarose, the most suitable agarose type for extrusion bioprinting can be obtained. Agarose solution can be rapidly gelled when the temperature drops to between 26 and 30 °C, which makes it suitable as a bioprintable material in scaffold fabrication. Living cells can be mixed into biocompatible agarose solutions, and agarose hydrogels are able to support the differentiation of encapsulated cells; however, other cellular functions such as biosynthesis of proteins and proteoglycans can be reduced compared to cells encapsulated in other protein-based biomaterials.

Because of its inert nature with respect to cell adhesion, agarose is also used as a nonadhesive hydrogel for the formation of cell aggregates. Also, agarose has been used as a "sacrifice biomaterial" in scaffold vascularization due to its thermosensitivity. In this strategy, agarose fibers are printed with a predefined pattern, with the functional biomaterials and cells then cast over the fibers and crosslinked. Using temperature control, the agarose fibers can be easily de-crosslinked as a solution and removed, with the patterned channels left behind.

*Hyaluronic acid (HA)* is a high molecular weight, natural linear polysaccharide of repeating β-D-glucuronic acid and N-acetyl-β-D glucosamine units. It is distributed throughout the human body and is predominantly involved in connective, epithelial, and neural tissues. HA has been extensively used in clinics as a dermal filler for wound healing and has lubricating properties as synovial fluid in articular joints.

HA is water soluble, with the resulting solution having a high viscosity. Therefore, it is often used as an assistant material to adjust the viscosity of other biomaterial solutions in bioprinting. For example, low-viscosity alginate solutions normally provide a soft environment after gelation that is suitable for cell performance, but printing these solutions in extrusion bioprinting is challenging if the viscosity is too low. By adding HA, the viscosity of the alginate solution can be regulated during bioprinting and the appropriate mechanical strength achieved.

Tunable physical and biological properties make HA a suitable material for incorporating cells. For example, HA is the major tissue extracellular matrix (ECM) component of cartilage, and therefore chondrocyte-encapsulated HA-based scaffolds have been fabricated with expected high cell viability and function. HA can gel by covalent crosslinking with hydrazide derivatives, by esterification, or by annealing, but all of these processes take a long time and can be toxic to encapsulated cells. Additionally, gelled HA has poor mechanical properties and is characterized by a rapid degradation rate. To address these shortcomings, HA is normally modified with UV-curable methacrylate (MA) to become photopolymerizable. The HA-MA maintains the crucial biological properties of HA while gaining controllable crosslinking properties, which greatly improves crosslinking efficacy and mechanical stability for scaffold bioprinting applications.

*Collagen* is an abundant, naturally occurring protein in the body that consists of self-aggregating polypeptide chains held together by both hydrogen and covalent bonds. It is the most widely used natural material for tissue scaffolds due to its natural receptors for cell attachment, creating the possibility to directly affect cell adhesion and other functions.

Some collagens are more compatible with bioprinting applications than others. The most widely used collagen formations in tissue engineering include collagen types I, II, IV, and V. Among these, collagen type I has been extensively applied in scaffold bioprinting. It is dissolvable in faintly acidic aqueous solutions and can be polymerized in 30–60 min at 37 °C at neutral pH. As such, collagen scaffolds can be printed by controlling the pH and temperature. In cell-incorporated bioprinting, a physiological environment with neutral pH is required, and thus the neutralization of collagen solution by an alkaline solution (e.g., NaOH) is necessary before cell mixing.

Collagen scaffolds have been used with diverse cell types, including adipose, bladder, blood vessel, bone, cartilage, heart, liver, nerve, and skin tissues, among many others. However, they face the limitation of inherently low mechanical properties. To make collagen more suitable for scaffold bioprinting and tissue engineering applications, covalent bonding and irradiation crosslinking methods have been applied along with the thermal polymerization of collagen solutions. Additionally, mixing collagen solution with other materials such as alginate, gelatin, and HA has also been adopted in scaffold bioprinting to improve mechanical properties. For the repair of many hard tissues, such as bone and cartilage, scaffolds made from collagen and synthetic polymers such as polycaprolactone (PCL) and poly(lactic-co-glycolic acid) (PLGA) have often been used. The synthetic polymers are printed first in a designated pattern as a scaffold frame to provide the mechanical support for the structures and cells, then collagen and cells are subsequently printed inside the spaces created by the frame to realize the biological functions of the scaffold.

*Gelatin* derived through partial hydrolysis of collagen has advantages such as good biocompatibility, nonimmunogenicity, and complete biodegradability in vivo [17]. Gelatin is widely used for tissue engineering applications as it possesses a similar composition to collagen. Analagous to collagen, gelatin is sensitive to temperature and crosslinked at low temperatures. When the temperature increases to the physiological range or higher, gelation de-crosslinks and shows instability. Therefore, for extrusion-based bioprinting applications, chemicals including metal ions, glutaraldehyde, or even other printable materials have been used to improve the printability and stability of gelatin. Photo-crosslinkable gelatin hydrogels have been synthesized by chemically modifying gelatin with methacrylamide side groups. Synthesized gelatin methacrylate composite (GelMA) hydrogels have been successfully printed through a pneumatic dispenser system supplemented with a UV light. These GelMA hydrogels have been employed to encapsulate various cell types for the fabrication of tissue-engineered cardiac valves, cartilage, and vessel-like structures. The mechanical properties of modified gelatin can be regulated by controlling the gelatin concentration, UV light intensity, or exposure time.

*Fibrin* is a fibrous protein that naturally forms in the body during blood coagulation. It contains fibrinogen, which is a protein type comprising two sets of three polypeptide chains: Aα, Bβ, and γ chains. Fibrinogen can be gelled to form fibrin hydrogel by adding thrombin, a serine protease that converts fibrinogen into fibrin. Coagulation factor XIII can covalently crosslink with the γ chains in the fibrin polymer to produce a fibrin network that is stable and resists protease degradation.

During bioprinting, fibrin can be simply achieved via direct deposition of fibrinogen solution into a mixture containing thrombin and factor XIII.

Fibrin-based scaffolds have an inherent cell adhesion capacity, which encourages many applications based on mixing cells into fibrinogen solutions to build cell-incorporated fibrin scaffolds. However, the utilization of fibrin scaffolds is limited by their low mechanical stability and rapid degradation. Methods to improve their mechanical properties include using high concentrations of fibrinogen or thrombin during fibrin formation or mixing fibrinogen with other biomaterials that provide better mechanical stability. The rapid degradation rate can be moderated by adding protease inhibitors such as aprotinin into the fibrinogen solutions or culture medium, or by optimizing the printing temperature, calcium ion concentration, and cell density.

Building fibrin-based scaffolds by extrusion bioprinting is challenging due to the limited viscosity of fibrinogen solution. Premixed fibrinogen and thrombin solutions have been applied in fibrin-based scaffold fabrication to improve the viscosity of printed solutions. Fibrinogen solution with a pre-determined ratio of fibrinogen and thrombin is normally prepared at a low temperature (around 0 °C) to moderate the gelation. Other methods to improve the printability of fibrin include mixing fibrinogen with other biomaterials during solution preparation and crosslinking them thereafter with associated crosslinkers for scaffold bioprinting.

*Decellularized matrix (dECM)* materials have drawn much attention for tissue engineering applications. Recent development methods for decellularizing tissue components not only provide opportunities for researchers to analyze the composition, localization, and biological functions of ECM but also create new biomaterial types for scaffold bioprinting [19, 20]. Work on tissue decellularization has attracted considerable attention for the regeneration of the heart, kidney, liver, and other organs. Decellularization is a process involving the lysis and removal of cellular components by the perfusion of deionized water or other mild detergents while preserving the tissue ECM. Because the ECM contains various molecules such as collagen, fibrin, and other proteins as well as growth factors that facilitate cell growth and functions, it has great advantages for biomimetic tissues and organs after careful tuning as a biomaterial for cell-incorporated bioprinting.

dECM components can be dissolved into an aqueous solution to meet extrusion property requirements, then solidified using temperature and pH control. dECM scaffolds have limited applications to hard tissue repair due to their weak mechanical properties, and thus other polymers such as PCL are normally adopted as frameworks to improve the structural stability. Because dECM has not been well studied and applied in tissue bioprinting, some limitations related to utilization protocols remain. Moreover, because dECM is obtained from natural organs and tissues, the amount of ECM remaining after decellularization is normally small and thus insufficient for the regeneration of large tissues. Toxic residuals during decellularization can remain in the dECM and reduce the viability and other functions of incorporated cells. As a result, more efforts are needed to address these various issues with dECM materials.

## 3.3.2   Synthetic Hydrogels

*Poly(ethylene)-based polymers* (mainly poly(ethylene glycol) (PEG) and poly(ethylene oxide) (PEO)) are the most widely used synthetic hydrogels in scaffold bioprinting for tissue engineering applications, primarily due to their tailorable properties. They are produced by the polymerization of ethylene oxide by condensation and can be classified as PEG or PEO based on molecular weight. PEG and PEO are hydrophilic, are compatible with reduced immunogenicity after implantation, and can be dissolved in water. The solutions can be crosslinked into hydrogels via physical, ionic, or covalent bonding methods.

PEG and PEO hydrogels possess high permeability that facilitates the exchange of nutrients and waste materials to support cell metabolism and are therefore often adopted to encapsulate cells for cell delivery. However, these hydrogels have limited protein binding and cell adhesion due to their inherent properties. To overcome this limitation, PEG/PEO hydrogels are often modified with peptides, such as RGD peptide, that have the capacity to enhance cell adhesion. For scaffold bioprinting, PEG/PEO solutions are often tailored to be photopolymerizable using either acrylates or methacrylates, with the modified PEG solution efficiently crosslinked by UV light to achieve improved mechanical stability after extrusion.

*Pluronic®* is a tri-block copolymer based on polyoxyethylene-polyoxypropylene-polyoxyethylene (PEO-PPO-PEO). It is sensitive to temperature because PPO side chains become less soluble above a threshold temperature between 22 and 37 °C (depending on polymer concentration) and gelation occurs. However, owing to its synthetic nature, Pluronic has disadvantages including limited cell adhesion and an inability to degrade. Previous research also shows it dissolves after 1 week of in vitro culture and has questionable cytocompatibility due to potential disruption of the cell membrane. Cell-laden Pluronic hydrogels show decreased viability and rapid dissolution after a few days of in vitro culture. Nonetheless, the advantages of Pluronic, such as high viscosity and good printability, still make it attractive for bioprinting constructs with good shape fidelity and accurate structures.

*Poly(ε-caprolactone)* (PCL) is a biodegradable polyester with a low melting point of 60 °C. It can be degraded by *hydrolysis* in physiological conditions (such as in the human body) and therefore has drawn considerable attention for use as an implantable *biomaterial*. Nowadays, PCL has been utilized in printing scaffolds for tissue engineering due to its low melting point (60 °C) readily for printing, as well as its thermoplastic behavior, respectable mechanical strength, and hydrolysis-induced biodegradation profile. In its melting phase, PCL is stable thermally and has a flow behavior appropriate for printing. On the printing stage with a low temperature, PCL can quickly solidify and form a mechanically stable 3D construct with good printability [Refs.]. Notably, the melting temperature of PCL (60 °C) is too high to sustain cell viability; therefore, printed PCL scaffolds require cell seeding after scaffold fabrication, or impregnation or printing with another hydrogel bioink with cells to form a hybrid structure of scaffolds [21]. As such, PCL is not considered as a biomaterial to directly incorporate cells, but rather an additive network or framework to provide hydrogel bioinks with a supportive structure.

### 3.3.3 Composite Hydrogels

Using composite biomaterials is another approach to mimic the physiological niche and improve the regeneration of tissue engineering outcomes. The close resemblance of natural or synthetic hydrogels to native tissues makes them attractive scaffold materials for soft tissue engineering. However, modulating the mechanical performance and degradation rate of single component materials can be difficult. Sometimes composite materials are needed, combining the superior properties of each component to mimic tissue mechanical properties as well as optimize the degradation rate.

There are generally two types of hydrogel composites based on their composition. The first composite hydrogel type consists of two or more hydrogel-forming polymers. For example, a composite hydrogel composed of collagen type I and an extracellular matrix protein was designed and showed improved mechanical and biological characteristics compared to gels obtained from the individual components [22]. Another study synthesized an alginate and collagen type I fibril composite hydrogel and showed that the addition of collagen type I fibrils improved the rheological and indentation properties of the resulting composite hydrogel [23]. Alginate and gelatin composite hydrogel constructs have also been bioprinted to fabricate living valve conduits with anatomical architecture and dual cell types that were viable over 7 days in culture [24]. The elastic modulus and mechanical strength of cell-laden hydrogels were maintained through 7 days of culture while the acellular printed hydrogels experienced a decrease in mechanical strength and modulus. These results demonstrate anatomically complex, heterogeneous cell-encapsulating aortic valve conduits can be bioprinted using composite alginate/gelatin hydrogels.

The second important type is inorganic filler-reinforced composite hydrogels. Most widely used inorganic fillers for hydrogels are inorganic ceramic-like hydroxyapatite [25] or carbon-based materials such as graphene. Inspired by the naturally occurring bioactive nanomaterials found in biological systems, researchers are developing novel bioactive biomaterials by combining inorganic ceramics with natural or synthetic polymers. A wide range of bioactive ceramic nanoparticles, including hydroxyapatite, silicate nanoparticles, and calcium phosphate, have been applied to synthesize composite hydrogels. A methylcellulose hydrogel containing calcium phosphate nanoparticles was prepared, with an in vitro study showing the prepared composite hydrogel was biocompatible. The in vivo study demonstrated the regeneration rate of newly formed bone was also higher in the composite hydrogel than in pure methylcellulose hydrogel [25]. The aim of incorporating calcium phosphate into a hydrogel is usually to enhance mechanical properties and introduce osteoinduction and bioactivity to the composite.

Carbon-based materials can also be introduced to increase the composite hydrogel system conductivity for biomedical engineering applications. For example, currently available biomaterials for cardiac tissue engineering lack electrical conductivity and appropriate mechanical properties, which are two parameters that play key roles in regulating cardiac cell behavior. A study that engineered myocardial tissue constructs based on graphene oxide and gelatin metacryloyl hybrid hydrogels showed the incorporation of graphene oxide into the gelatin metacryloyl matrix significantly enhanced the electrical conductivity and mechanical properties of the resulting composite hydrogels [26]. Embedded cardiac cells in the composite hydrogel scaffolds demonstrate superior biological activities in terms of cell viability, proliferation, and differentiation compared to those cultured in pure gelatin metacryloyl hydrogels [27]. In other work, carbon nanotubes were combined with collagen type I hydrogels to characterize potential improvements in hydrogel strength and conductivity. Cardiomyocytes seeded in the carbon nanotubes-collagen hybrid hydrogels showed improved cardiac cell functions compared to those within pure collagen hydrogels. Graphene and graphene oxide have also been mixed with polyurethane to prepare graphene-based nanocomposite hydrogel bioinks for bioprinting. The rheological properties of the graphene-based composite hydrogel were suitable for the printing and survival of neural stem cells. Numerous research works in this area suggest the great promise of carbon-based materials as a functional component of bioprinted or engineered scaffolds [14, 28].

## 3.4  Summary

A bioink for bioprinting is a formulation of biomaterial(s), living cells, and/or bioactive molecules and its properties are important in bioprinting scaffolds for tissue engineering. Printability, crosslinking mechanisms, biocompatibility, biodegradation, and mechanical properties are important factors in consideration when formulating bioinks.

The printability of a bioink refers to its ability to form and maintain a 3D structure by printing. To characterize the bioink prointabilty, the common practice is to examine its extrudability, filament fidelity, and structural integrity. The printability of a bioink can be affected by many factors and among them, the flow behavior (viscosity) and crosslinking mechanisms are the most important. Typically, higher viscosity will result in better printability but in turn negatively affect cell behavior. Potential crosslinking mechanisms for bioprinting include ionic, thermal, and photocrosslinking. The easier and quicker the crosslinking process, the better printability the hydrogel-forming solutions will have.

Hydrogels can be degraded via hydrolysis and/or enzymatically induced cleavage. Biodegradation through hydrolysis occurs in the forms of bulk and/or surface degradation; while in enzymatic degradation, the bond/chains cleavage in the hydrogel is caused by the catalytic action of enzymes and its rate can be regulated by the enzyme concentration.

Hydrogel-forming biomaterials can be classified as naturally derived or synthetic. Most hydrogels used for formulating bioinks are derived from natural extracellular matrix sources, including collagen, gelatin, fibrin, HA, alginate, and agarose. These naturally derived polymers have intrinsic capabilities to support cell viability and proliferation and can be relatively easily degraded or metabolized. However, the limitations of natural biomaterials include rejection and/or immune-related sequelae, quick degradation rates, and poor mechanical properties. Synthetic hydrogels are usually biologically inert and nonbiodegradable, but are open to modification to overcome the disadvantages of natural hydrogels. Due to the complexity of targeted tissue, more than one type of biomaterial is usually needed to formulate a bioink. Naturally derived polymers potentially support cell viability and proliferation and can be degraded. Their limitations include rapid degradation rates and poor mechanical properties. Although synthetic hydrogels are usually biologically inert and nonbiodegradable, their use can potentially overcome the abovementioned disadvantages of natural hydrogels.

**Problems**
1. What are the important properties of a biomaterial or bioink for bioprinting?
2. What is the printability of a bioink? Explain how the printability is affected by the bioink properties.
3. Briefly explain biocompatibility and biodegradation and why they are important in bioprinting.
4. Briefly explain the crosslinking mechanisms that are commonly used in bioprinting.
5. Name and briefly explain one of the hydrogel biodegradation mechanisms.
6. How can the hydrogels used in bioprinting be classified? Briefly explain the merits and demerits of each class of hydrogels.
7. Choose a commonly used hydrogel and briefly explain its use in bioprinting.
8. Name and explain one purpose of incorporating carbon-based materials into composite biomaterials.

# References


1. A. Malekpour, X. Chen, Printability and cell viability in extrusion-based bioprinting from experimental, computational, and machine learning views. J. Funct. Biomater. **2022**(13), 40 (2022)
2. Z.Q. Fu, S. Naghieh, C.C. Xu, C.J. Wang, W. Sun, X.B. Chen, Printability in extrusion bioprinting. Biofabrication **13**(3), 033001 (2021)
3. S. Naghieh, X.B. Chen, Printability – A key issue in extrusion-based bioprinting. J. Pharm. Anal. **11**(5), 564–579 (2021)
4. A. Zimmerling, Z. Yazdanpanah, D. Cooper, J. Johnston, X.B. Chen, 3D printing PCL/nHA bone scaffolds: Exploring the influence of material synthesis techniques. Biomater. Res. **25**, 3 (2021)

5. N. Soltan, L.Q. Ning, F. Mohabatpour, P. Papagerakis, X.B. Chen, Printability and cell viability in bioprinting alginate dialdehyde-gelatin scaffolds. ACS Biomater Sci. Eng. **5**(6), 2976–2987 (2019)

6. A. Rajaram, D.J. Schreyer, X.B. Chen, Use of the polycation polyethyleneimine to improve the physical properties of alginate-hyaluronic acid hydrogel during fabrication of tissue repair scaffolds. J. Biomater. Sci. Polym. Ed. **26**(7), 433–445 (2015)

7. A. Rajaram, D.J. Schreyer, X.B. Chen, Bioplotting alginate-hyaluronic acid hydrogel scaffolds with structural integrity and preserved Schwann cell viability. 3D Print. Addit. Manuf. **1**(4), 194–203 (2014)

8. F. You, X. Wu, X. Chen, 3D printing of porous alginate/gelatin hydrogel scaffolds and their mechanical property characterization. Int. J. Polym. Mater. Polym. Biomater. **66**(6), 299–306 (2017)

9. M.G. Li, X.Y. Tian, N. Zhu, D. Schreyer, X.B. Chen, Modeling process-induced cell damage in the bio-dispensing process. Tissue Eng. Part C **16**(3), 533–542 (2009)

10. L.Q. Ning, B. Yong, F. Mohabatpour, B. Nicholas, M.D. Sarker, P. Papagerakis, X.B. Chen, Process-induced cell damage: Pneumatic vs. screw-driven bioprinting. Biofabrication **12**(3), 025011 (2020)

11. M.D. Sarker, M. Izadifar, D.J. Schreyer, X.B. Chen, Influence of ionic crosslinkers (Ca$^{2+}$/ Ba$^{2+}$/ Zn$^{2+}$) on the mechanical and biological properties of 3D bioplotted hydrogel scaffolds. J. Biomater. Sci. Polym. Ed. **29**, 1126–1154 (2018)

12. S. Naghieh, M.D. Sarker, E. Karki, M.R. Karamooz-Ravari, X.B. Chen, Influence of crosslinking on the mechanical behavior of 3D printed alginate scaffolds: Experimental and numerical approaches. J. Mech. Behav. Biomed. Mater. **80**, 111–118 (2018)

13. L. Sando, S. Danon, A.G. Brownlee, et al., Photochemically crosslinked matrices of gelatin and fibrinogen promote rapid cell proliferation. J. Tissue Eng. Regen. Med. **5**, 337–346 (2011)

14. M. Izadifar, D. Chapman, P. Babyn, X.B. Chen, M. Kelly, UV-assisted 3D bioprinting of nano-reinforced hybrid cardiac patch for myocardial tissue engineering. Tissue Eng. Part C Methods **24**(2), 74–88 (2018)

15. L.Q. Ning, N. Betancourt, D. Schreyer, X.B. Chen, Characterization of cell damage and proliferative ability during and after bioprinting. ACS Biomater Sci. Eng. **4**(11), 3906–3918 (2018)

16. H.R. Bakhsheshi-Rad, Z. Hadisi, A.F. Ismail, M. Aziz, M. Akbari, F. Berto, X.B. Chen, In vitro and in vivo evaluation of chitosan-alginate/gentamicin wound dressing nanofibrous with high antibacterial performance. Polym. Test. **82**, 106298 (2020)

17. A. Sadeghianmaryan, S. Naghieh, H.A. Sardroud, Z. Yazdanpanah, Y.A. Soltani, J. Sernagli, X.B. Chen, Extrusion-based printing of chitosan scaffolds and their in vitro characterization for cartilage tissue engineering. Int. J. Biol. Macromol. **164**(1), 3179–3192 (2020)

18. A. Sadeghianmaryan, S. Naghieh, Z. Yazdanpanah, H.A. Sardroud, N.K. Sharma, L.D. Wilson, X. Chen, Fabrication of chitosan/alginate/hydroxyapatite hybrid scaffolds using 3D printing and impregnating techniques for potential cartilage regeneration. Int. J. Biol. Macromol. **204**, 62–75 (2022)

19. A. Fazel Anvari-Yazdi, K. Tahermanesh, D.J. MacPhee, I. Badea, M. Ejlali, A. Babaei-Ghazvini, B. Acharya, X. Chen, Comparative Analysis of Porcine-Uterine Decellularization for Bioactive-Molecules Preservation and DNA Removal. Frontiers in Bioengineering and Biotechnology, 12, p.1418034 (2024). https://www.frontiersin.org/journals/bioengineering-and-biotechnology/articles/10.3389/fbioe.2024.1418034/full

20. A.D. McInnes, M.A.J. Moser, X. Chen, Preparation and use of Decellularized extracellular matrix for tissue engineering. J. Funct. Biomater. **13**(4), 220 (2022)

21. Z. Izadifar, T.J. Chang, W. Kulyk, X.B. Chen, B.F. Eames, Analyzing biological performance of 3D-printed, cell-impregnated hybrid constructs for cartilage tissue engineering. Tissue Eng. Part C **22**(3), 173–188 (2015)

22. M. Maisani, S. Ziane, C. Ehret, et al., A new composite hydrogel combining the biological properties of collagen with the mechanical properties of a supramolecular scaffold for bone tissue engineering. J. Tissue Eng. Regen. Med. **12**(3), e1489–e1500 (2018)
23. M. Baniasadi, M. Minary-Jolandan, Alginate-collagen fibril composite hydrogel. Materials (Basel) **8**(2), 799–814 (2015)
24. B. Duan, L.A. Hockaday, K.H. Kang, J.T. Butcher, 3D bioprinting of heterogeneous aortic valve conduits with alginate/gelatin hydrogels. J. Biomed. Mater. Res. A (5), 101A, 1255–1264A (2013)
25. M.H. Kim, B.S. Kim, H. Park, J. Lee, W.H. Park, Injectable methylcellulose hydrogel containing calcium phosphate nanoparticles for bone regeneration. Int. J. Biol. Macromol. **109**, 57–64 (2018)
26. J. Zhou, J. Chen, H. Sun, et al., Engineering the heart: Evaluation of conductive nanomaterials for improving implant integration and cardiac function. Sci. Rep. **4**, 3733 (2014)
27. S.R. Shin, C. Zihlmann, M. Akbari, et al., Reduced graphene oxide-gelMA hybrid hydrogels as scaffolds for cardiac tissue engineering. Small **12**(27), 3677–3689 (2016)
28. C.-T. Huang, L. Kumar Shrestha, K. Ariga, S. Hsu, A graphene–polyurethane composite hydrogel as a potential bioink for 3D bioprinting and differentiation of neural stem cells. J. Mater. Chem. B **5**, 8854–8864 (2017)

# Chapter 4
# Mechanical Properties of Native Tissues and Scaffolds

## 4.1 Introduction

Tissue engineering scaffolds, once fabricated, are typically incubated in bioreactors (in vitro) to mature before they are subsequently implanted within living bodies (in vivo). Tissue scaffolds provide cells with a 3D structure and mechanical support for cellular processes such as migration, proliferation, and/or differentiation, thus facilitating the growth of a functional tissue-engineered construct or viable tissue. During this process, the mechanical properties of scaffolds play a crucial role. This chapter focuses on common testing methods, including those for tensile/compressive, bending, torsion, creep, relaxation, cyclic load, and frequency-dependent aspects, to measure and characterize the mechanical (both static and dynamic) properties of native tissues and scaffolds. This chapter also presents general considerations for mechanical property measurements, case studies for measuring the mechanical properties of native tissues and scaffolds, and methods to improve the mechanical properties of scaffolds.

## 4.2 Mechanical Testing Methods

### 4.2.1 Basics of Mechanical Testing

Mechanical characterization of both native tissues and scaffolds is typically performed on a mechanical testing instrument or system, like the one shown in Fig. 4.1. During mechanical testing, the specimen or sample of native tissue or scaffold mounted in the test chamber is subjected to loading conditions according to the type of testing while the deformation of the specimen is measured or recorded. The loading conditions commonly applied to specimens during mechanical testing are shown in Fig. 4.2, including tensile/compressive force ($F$), bending moment


D. X. B. Chen, *Extrusion Bioprinting of Scaffolds for Tissue Engineering*, https://doi.org/10.1007/978-3-031-72471-8_4

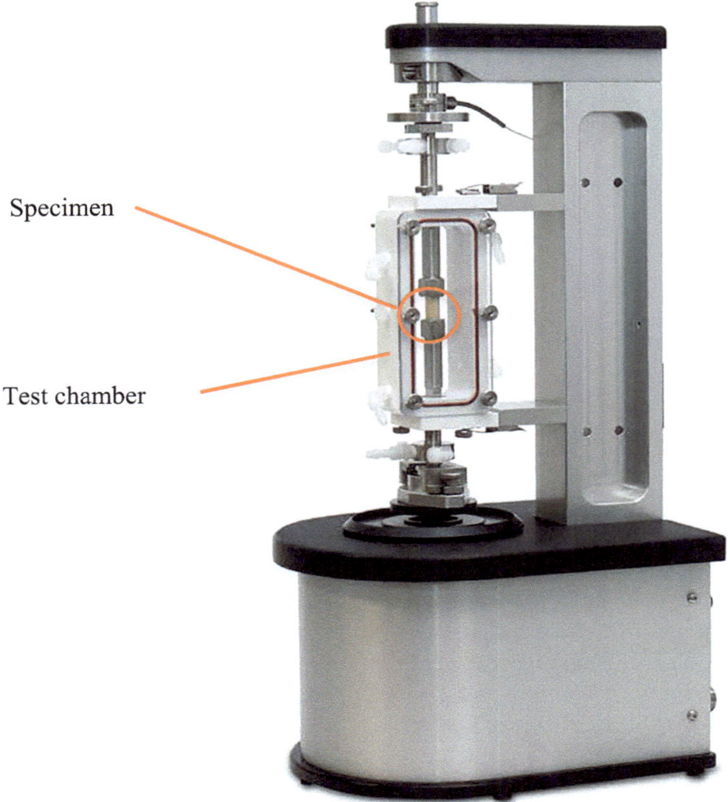


**Fig. 4.1** A mechanical testing instrument for determining the mechanical properties of native tissues and scaffolds

($M$), and torque ($T$), along with the deformation caused. A specimen subjected to tensile force is elongated in the axial direction (Fig. 4.2a), whereas a specimen subjected to compressive force is compressed along the axial direction (Fig. 4.2b). A specimen subjected to a bending moment is deformed with the upper part being elongated and lower part being compressed (Fig. 4.2c). A specimen subjected to torsion is twisted by an angle of $\varphi$ (Fig. 4.2d).

In mechanical testing, stress and strain are two important concepts associated with the force applied and the deformation caused, respectively. Consider a cylindrical specimen subjected to axial tensile loading as illustrated in Fig. 4.3a. As this specimen is deformed by external applied loading, internal resistive forces develop inside the material. These internal forces per unit area are called stresses. The even distribution of resistive forces over an arbitrary cross section of the cylindrical specimen is shown in Fig. 4.3b. The stress developed due to tensile loading is called *normal stress*, as it develops perpendicular to the cross section of the specimen. If

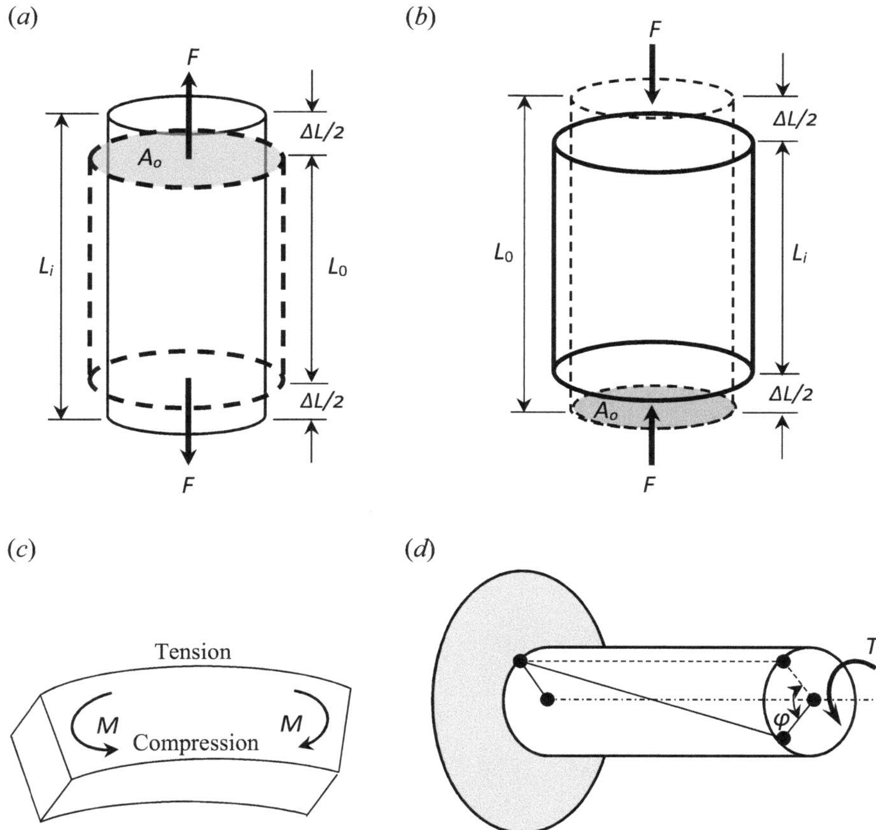



**Fig. 4.2** Types of loading conditions applied to a specimen: (**a**) tensile force ($F$), (**b**) compressive force ($F$), (**c**) bending moment ($M$), and (**d**) torsion ($T$). $L_0$ is the original length, $L_i$ is the instantaneous length, $A_0$ is the original area, $\Delta L$ is the change in length, and $\varphi$ is the angular deformation

one reverses the direction of loading, opposing stresses will also develop to resist shortening of the specimen. This type of normal stress is called *compressive stress*. Normal stress is typically denoted by the lowercase Greek letter sigma ($\sigma$) as:

$$\sigma = \frac{\text{Applied force } (F)}{\text{Cross} - \text{sectional area } (A)}. \tag{4.1}$$

The unit of stress is Newton's per meter squared ($N/m^2$), also known as Pascal (i.e., Pa).

The other type of stress is called *shear stress*, which develops inside a specimen if a tangential load is applied to produce shear deformation. Shear (tangential) loading

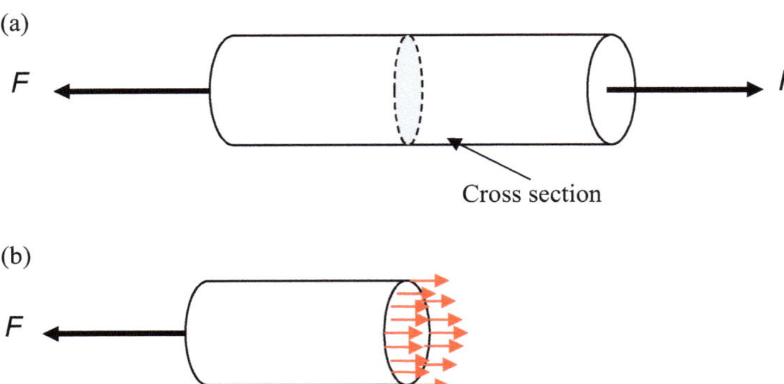



**Fig. 4.3** (**a**) Diagram of a cylindrical specimen subjected to axial tensile loading. (**b**) Free body diagram showing the distribution of internal resistive forces over the specimen cross section

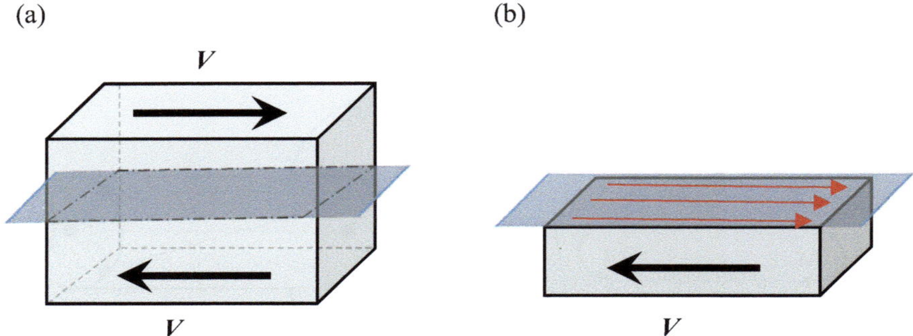



**Fig. 4.4** (**a**) Diagram showing the tangential shear load ($V$) applied to a cuboid-shaped specimen. (**b**) Free body diagram of the specimen showing the even distribution of shear resistive forces over the cross-sectional area

applied to a cuboid-shaped specimen is shown in Fig. 4.4a, with the even distribution of shear stresses over the cross-sectional area of the specimen shown in Fig. 4.4b. Shear stress is denoted by the lowercase Greek letter tau ($\tau$) as:

$$\text{Shear stress } (\tau) = \frac{\text{Tangentially applied load } (V)}{\text{Cross} - \text{sectional area } (A)}. \qquad (4.2)$$

Strain is associated with the deformation of a specimen due to external applied loading. Consider the case of elongation of a specimen due to externally applied tensile loading. The total elongation of the specimen is designated by $\Delta L$. The average extensional strain in the specimen is defined as the ratio of total elongation to original length $L_o$. The *normal strain* is represented by the lowercase Greek letter epsilon ($\epsilon$), and average normal tensile strain can be defined as:

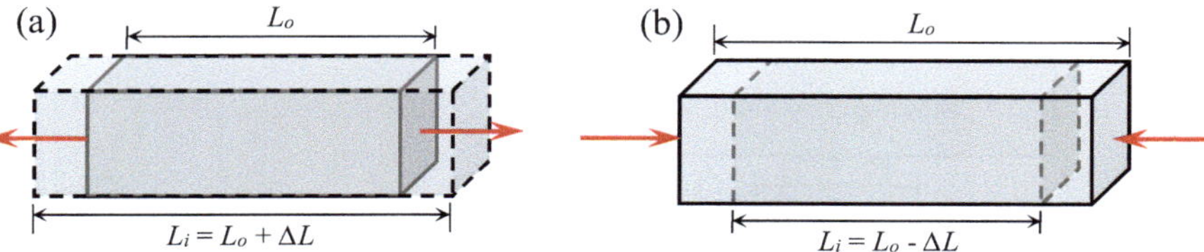

**Fig. 4.5**  Deformation of a specimen under axial (**a**) tensile or (**b**) compressive loading

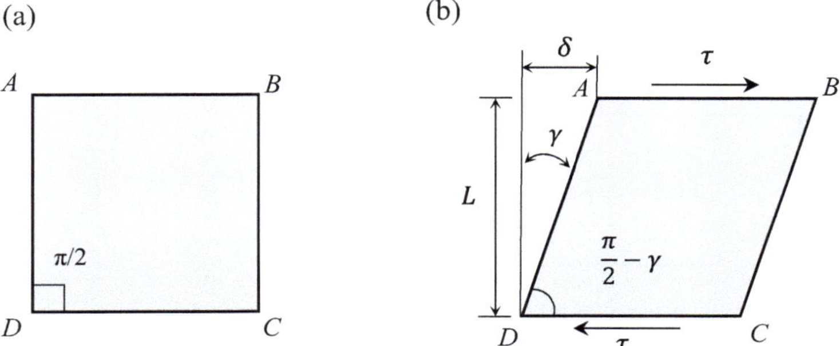

**Fig. 4.6**  Diagrams showing the (**a**) original shape of a rectangular element and (**b**) deformed shape under pure shear

$$\epsilon_{avg} = \frac{\Delta L}{L} = \frac{L_i - L_o}{L_o}, \tag{4.3}$$

where $L_i$ is the length of the specimen after application of the load as shown in Fig. 4.5a. On the other hand, compressive loading causes shortening of the specimen, as illustrated in Fig. 4.5b, and therefore results in a negative value for normal strain (because $L_i < L_o$ in this case), which is called normal compressive strain or simply *compressive strain*.

The deformation of a specimen under shear loading is different than for a specimen subjected to axial (tensile or compressive) loading. Specimens under shear loading have no tendency to elongate or shorten but instead deform or change in shape. The deformed shape of a rectangular element under shear stress is a parallelogram, as shown in Fig. 4.6, where the angle ($\gamma$) is used as a measure of the distortion or change in shape of the element and is called *shear strain*. The shear strain has units of degrees or radians.

For a small angular deformation, shear strain can be approximated by:

$$\gamma \approx \tan \gamma = \frac{\delta}{L}, \tag{4.4}$$

where $\delta$ and $L$ are the linear deformation and element length, as defined in Fig. 4.6b.

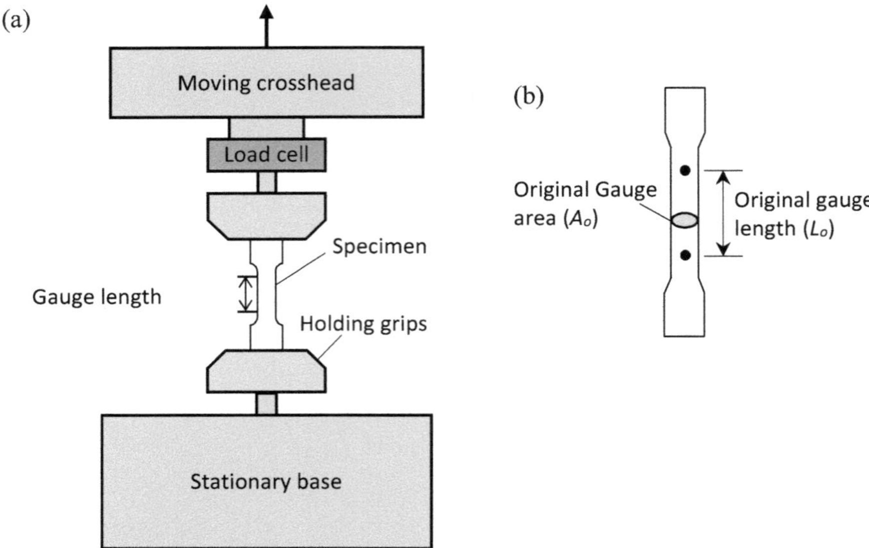


**Fig. 4.7**  Diagrams showing a (**a**) tensile test apparatus and (**b**) typical tensile specimen

## 4.2.2  *Tensile and Compressive Tests*

Tensile and compressive tests are performed to assess specimen behavior under uniaxial loading. These mechanical tests can be performed on a machine or apparatus, as schematically shown in Fig. 4.7a, where a tensile or compressive load is applied and monitored by a load cell and the corresponding deformation of the specimen due to applied load is measured by an extensometer or strain gauges on the specimen. Tensile tests are typically performed on a dumbbell-shaped specimen, as schematically shown Fig. 4.7b, where gauge points are marked to define the initial gauge length ($L_0$) and gauge area ($A_0$) of the specimen. Compressive tests are usually performed on cylindrical or cuboidal specimens.

A stress-strain curve is generated and recorded when a specimen is subjected to either tensile or compressive loading on a test apparatus. Figure 4.8 shows the stress-strain curves of common materials, including brittle materials such as ceramics, metals, and polymers.

A number of mechanical property parameters can be defined and determined using the recorded stress-strain curves. As an example, Fig. 4.9 shows a stress-strain curve along with some defined parameters.

The slope of the linear section of the curve gives the value of Young's modulus of elasticity (or elastic modulus), $E$, which is a measure of material stiffness. With further deformation, the material begins to yield or start plastic deformation, and the stress corresponding to the beginning of plastic deformation (or the end of elastic deformation) is known as the yield stress $\sigma_{\text{yield}}$ and the strain in the sample is the

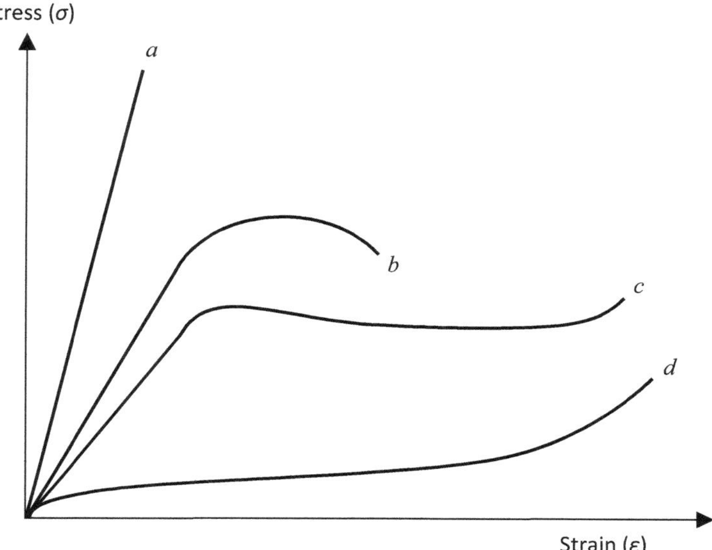


**Fig. 4.8** Diagram showing a range of stress-strain curves for different materials: (**a**) brittle materials such as ceramics, (**b**) metals, (**c**) hard and/or tough polymers, and (**d**) viscoelastic polymers

yield strain $\varepsilon_{yield}$. Notably, the transition from elastic to plastic deformation may not be readily apparent for some tissues and/or materials. In this case, it is common to use a 0.2% strain offset to determine the yield point or stress. Briefly, an offset line parallel to the linear portion of the stress-strain curve is drawn, beginning at a strain offset of 0.2% (or $\varepsilon_{offset}$), and the intersection of this offset line and the stress-strain curve is the yielding point, as shown in Fig. 4.9. Before yielding, the original geometry of the sample is recoverable upon the load removal. After yielding, the deformation imposed on the sample becomes permanent. After yielding fully, the sample continues to deform plastically until it finally reaches the ultimate stress ($\sigma_{ultimate}$) it can bear. After reaching the value of ultimate stress, the sample starts deforming very fast and finally fails at a stress of $\sigma_{failure}$ and a strain of $\varepsilon_{failure}$. The elastic modulus, yield strength, ultimate strength, and failure strength are, respectively, defined by:

$$E = \frac{\sigma}{\varepsilon}, \tag{4.5}$$

$$\sigma_{yield} = \frac{F_{yield}}{A_o}, \tag{4.6}$$

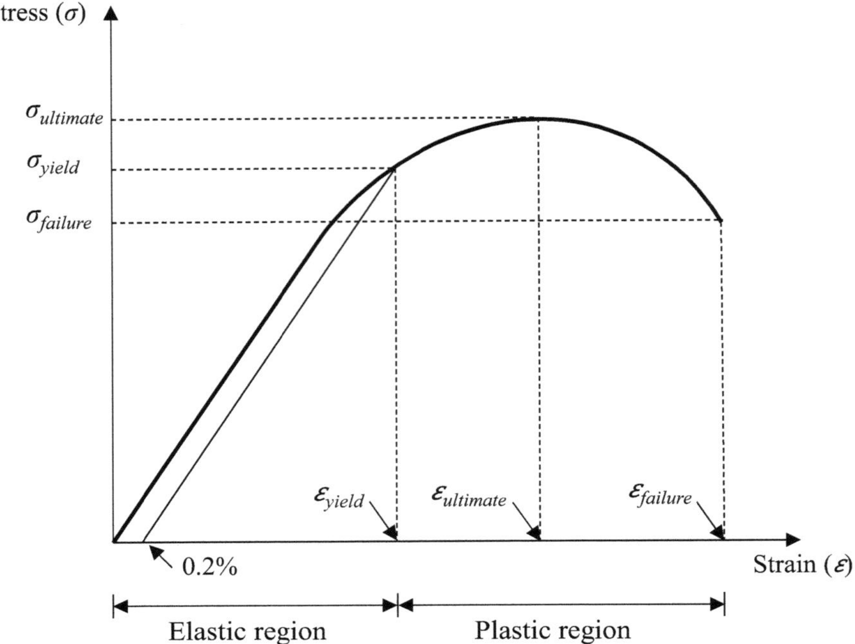


**Fig. 4.9** Representative stress-strain curve and defined parameters

$$\sigma_{\text{ultimate}} = \frac{F_{\text{ultimate}}}{A_o}, \tag{4.7}$$

and

$$\sigma_{\text{failure}} = \frac{F_{\text{failure}}}{A_o}. \tag{4.8}$$

The stress-strain diagram shown in Fig. 4.9 and the terms discussed above are all based on *engineering stress* ($\sigma$, or the load divided by the original gauge area) and *engineering strain* ($\varepsilon$, or the change in length divided by original length). The values of stress and strain evaluated using instantaneous parameters are called true stress and true strain. *True stress* ($\sigma_t$) is defined as the load applied at some instant divided by the instantaneous cross-sectional area of the specimen. *True strain* ($\varepsilon_t$) is the ratio of the instantaneous change in the length of the specimen to the instantaneous length of the specimen. True strain and true stress can be evaluated from the corresponding engineering stress and engineering strain using Eqs. (4.9) and (4.10), respectively:

$$\sigma_t = \sigma(1 + \varepsilon), \tag{4.9}$$

$$\varepsilon_t = \ln(1 + \varepsilon).  \tag{4.10}$$

Note that when an object is loaded and begins to deform, some of the applied energy is stored within the material while the remainder is dissipated as heat. The amount of energy dissipated is associated with the material damping and the rate at which strain is applied to the sample. In static testing, the rate of deformation should be sufficiently slow so that the effect of damping can be minimized or neglected.

## Example 4.1

A specimen of bovine tibiae cortical bone was tested under tensile loading. This specimen was designed as a strip-type dumbbell shape having a thickness of 2.5 mm, gauge width of 4 mm, gauge length of 25 mm, and total length of 60 mm. Tensile testing was performed at a rate of 1.8 mm/min, and the results of this test in terms of stress and strain are reported in the table below. Evaluate the following mechanical properties: (a) elastic modulus, (b) yield strength, (c) ultimate strength, and (d) failure strength.

| Strain | Stress (MPa) |
| --- | --- |
| 0 | 0 |
| 0.0004 | 12.54 |
| 0.0009 | 29.57 |
| 0.0013 | 41.93 |
| 0.0018 | 55.2 |
| 0.0023 | 70.97 |
| 0.0032 | 88.17 |
| 0.0043 | 102.87 |
| 0.0055 | 108.25 |
| 0.0069 | 114.35 |
| 0.009 | 118.67 |
| 0.0108 | 121.54 |
| 0.0127 | 123.7 |
| 0.0146 | 126.58 |
| 0.0162 | 128.38 |
| 0.0179 | 130.54 |
| 0.0202 | 133.42 |
| 0.023 | 137.74 |
| 0.0254 | 141.34 |
| 0.0281 | 144.58 |

## Solution

(a) The elastic modulus is the slope of the linear portion of the stress-strain curve. Three sets of data are selected from the linear portion of the stress-strain curve to determine the average value of the elastic modulus. The values of stress (in MPa) and strain corresponding to these sets are as follows: $\sigma_1 = 12.54$, $\varepsilon_1 = 0.0004$; $\sigma_2 = 29.57$, $\varepsilon_2 = 0.0009$; and $\sigma_3 = 41.93$, $\varepsilon_3 = 0.0013$.

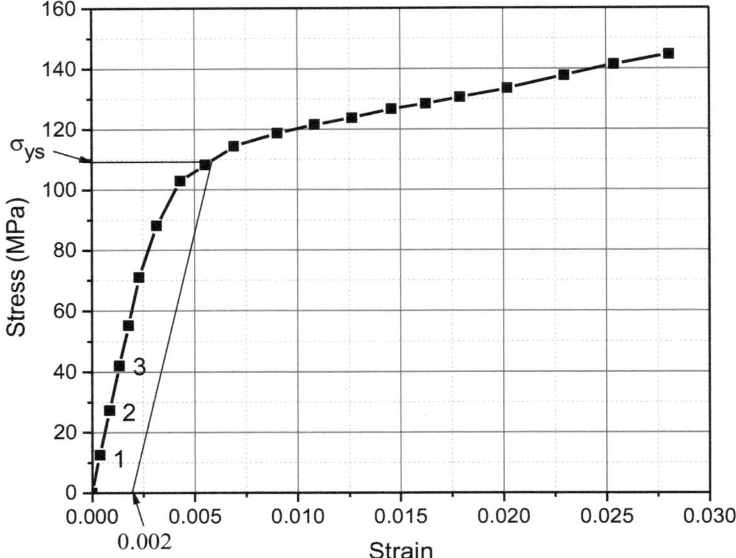


Stress-strain curve drawn from the data provided.

The elastic modulus values for these sets are calculated as follows:

$$E_I = \frac{\sigma_2 - \sigma_1}{\varepsilon_2 - \varepsilon_1} = \frac{27.24 - 12.54}{0.0009 - 0.0004} = 29400 \text{ MPa} = 29.4 \text{ GPa}$$

$$E_{II} = \frac{\sigma_3 - \sigma_2}{\varepsilon_3 - \varepsilon_2} = \frac{41.93 - 29.57}{0.0013 - 0.0009} = 30900 \text{ MPa} = 30.9 \text{ GPa}$$

$$E_{III} = \frac{\sigma_3 - \sigma_1}{\varepsilon_3 - \varepsilon_1} = \frac{41.93 - 12.54}{0.0013 - 0.0004} = 32655.5 \text{ MPa} = 32.6 \text{ GPa}$$

Average value of elastic modulus:

$$E = \frac{E_I + E_{II} + E_{III}}{3} = \frac{29.4 + 30.9 + 32.6}{3} = \mathbf{30.9 \text{ GPa}}$$

The elastic modulus value can also be determined by fitting the linear portion of stress-strain curve to a straight line, with a slope giving the value of elastic modulus. In the given example, the stress-strain curve of bone material is linear up to the stress value of 70.97 MPa, and the fitting results are shown in the following figure. The slope of this fitted line, or the elastic modulus, is 30,740 MPa or **30.74 GPa**.

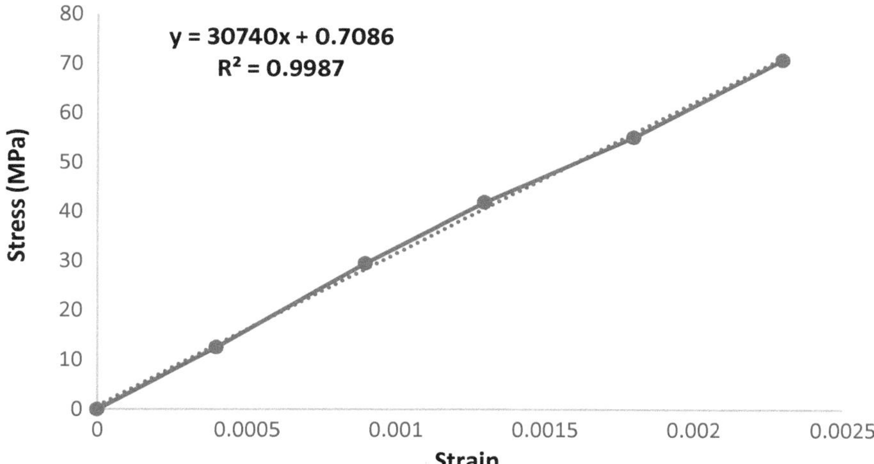


(b) The yield strength can be calculated using the 0.2% offset point method. A secant line with a slope equal to the initial slope of the stress-strain curve has been drawn from a point corresponding to a strain value of 0.002. The yield strength can be determined from the intersection point of this secant line and the stress-strain curve. The stress corresponding to this intersection point will be equal to the yield strength of the material. Therefore, yield strength $(\sigma_{ys}) =$ **108.3 MPa.**

(c) The ultimate strength is equal to the maximum value of stress obtained from the tensile test. The ultimate strength for the given material is therefore **144.58 MPa**.

(d) The failure strength is equal to the stress at which the material fails. In the given case, ultimate strength and failure strength are the same, that is, **144.58 MPa.**

### *4.2.3   Bending Tests*

The mechanical properties determined from bending tests, such as flexural elastic modulus ($E_f$), flexural strength ($\sigma_f$), and flexural strain ($\varepsilon_f$), are important to determine for tissues or scaffolds that are subjected to bending loads in their service environment. A bending test is performed by using a special test fixture as shown in Fig. 4.10. Typically, a rectangular or circular cross-section specimen (beam) with a length about 16 times its depth (thickness or diameter) is used in a three-point bending test.

The bending properties of a specimen subjected to a three-point bending test are evaluated using the following equations:

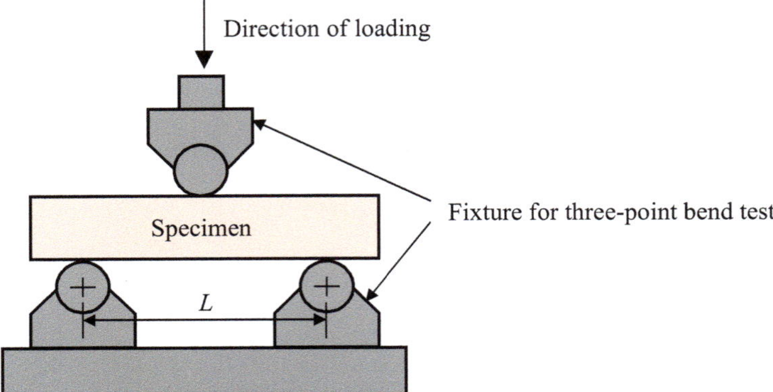


**Fig. 4.10** Schematic representation of a three-point bending test

$$\sigma_f = \frac{3FL}{2bd^2} \tag{4.11}$$

determines the flexural strength for a specimen with a rectangular cross section;

$$\sigma_f = \frac{FL}{\pi R^3} \tag{4.12}$$

determines the flexural strength for a specimen with a circular cross section;

$$\varepsilon_f = \frac{6d\delta_{max}}{L^2} \tag{4.13}$$

determines the flexural strain; and

$$E_f = \frac{L^3 m}{4bd^3} \tag{4.14}$$

determines the flexural elastic modulus, where $F$ is the maximum load sustained by the specimen, $L$ is the span length (as shown in Fig. 4.10), $b$ is the width of the specimen, $d$ is the thickness of the specimen, $\delta_{max}$ is the maximum deflection of the center point of the specimen, $m$ is the stiffness (slope) obtained from the initial linear portion of the load-deflection curve, and $R$ is the radius of a specimen with a circular cross section.

**Example 4.2**

A human femur bone was subjected to a three-point bending test with a fixture of span length 40 mm. The strip-type specimen for the bending test was prepared with a rectangular cross section and dimensions of $2 \times 5 \times 50$ mm$^3$. The maximum load recorded was 60.32 N and the maximum deflection of the bone was 0.38 mm. The initial data points recorded in terms of load and deflection values are reported in the table below. Calculate the (a) flexural strength, (b) elastic modulus, and (c) flexural strain of the cortical bone.

| Displacement (mm) | Load (N) |
| --- | --- |
| 0.00 | 0.00 |
| 0.01 | 10.00 |
| 0.05 | 11.55 |
| 0.10 | 13.49 |

**Solution**

Given: Span length $L = 40$ mm, thickness of specimen $d = 2$ mm, width of specimen $b = 5$ mm, maximum flexural load $F = 60.32$ N, and maximum deflection $\delta_{\max} = 0.38$.

(a) Flexural strength of the bone:

Flexural strength can be calculated from $\sigma_f = \frac{3FL}{2bd^2}$.

Therefore, $\sigma_f = \frac{3 \times 60.32 \times 40}{2 \times 5 \times 2^2} = \boldsymbol{180.96\ MPa.}$

(b) Elastic modulus of the bone:

The elastic modulus can be calculated from the initial slope ($m$) of the load-displacement curve using the relation $E_f = \frac{L^3 m}{4bd^3}$ .

The slope of the initial portion $m$ can be determined from the load-displacement values given in the table.

$$m = \frac{11.55 - 10.00}{0.05 - 0.01} = \frac{1.55}{0.04} = 38.75\ N/mm$$

Therefore, $E_f = \frac{40^3 \times 38.75}{4 \times 5 \times 2^3} = \frac{2480000}{160} = 15500\ MPa = \boldsymbol{15.5\ GPa.}$

(c) Flexural strain of the bone:

Flexural strain can be determined using the following equation:

$$\varepsilon_f = \frac{6d\delta_{\max}}{L^2} = \frac{6 \times 2 \times 0.38}{40^2} = \boldsymbol{0.00285\ mm/mm}$$

## 4.2.4   Torsion Tests

A torsion test is employed to determine the deformational behavior of a specimen subjected to a twisting moment. The twisting moment is applied to a specimen to twist it to a specified degree, with a specified torque, or until it fails under torsion. During this test, a cylindrical specimen is subjected to a twisting moment (torque)

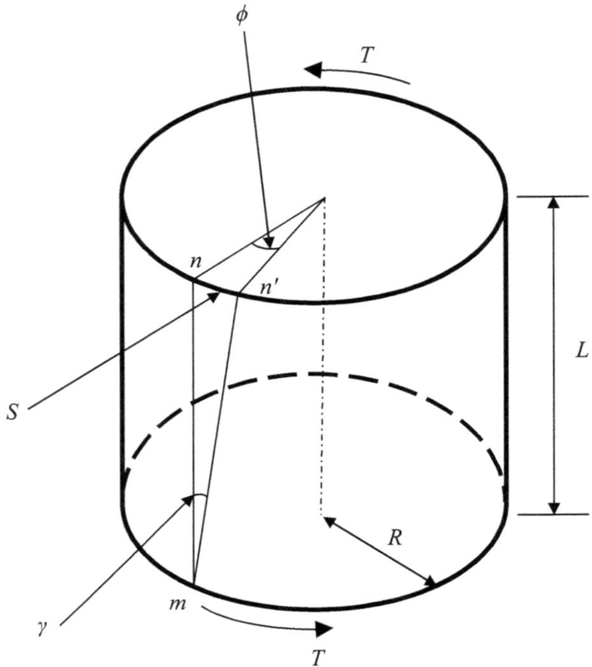

**Fig. 4.11** Twisting of a cylindrical sample under torsional loading

such that one end of the specimen remains fixed while the other end rotates about its axis with respect to the fixed end. The twisting moment may also be applied to both ends such that they rotate in opposite directions with respect to each other. Due to the twisting moment or torsional force applied, shear stresses and shear strains are induced inside the specimen. A cylindrical specimen subjected to a torsional load is shown in Fig. 4.11. Under the torque ($T$) applied to the cylindrical specimen, its upper surface rotates over an angle $\phi$ with respect to the bottom surface. Due to this rotation, a straight longitudinal line $mn$ on the surface of the specimen becomes a helical line $mn'$. The specimen deformation is considered as pure shearing, with a strain defined by $\gamma$. The magnitude of shear stress ($\tau$) can be determined from the shear strain using Hook's law:

$$\tau = G\gamma, \tag{4.15}$$

where $G$ is the shear modulus of elasticity. The relationship between the angle of twist $\phi$ and the shear strain $\gamma$ developed on the outer surface of the specimen is given by:

$$\gamma = \frac{R\varnothing}{L}, \tag{4.16}$$

where $R$ is the radius of the specimen and $L$ is the length as shown in Fig. 4.11. This equation can also be used to determine the values of shear strain inside the specimen (at radii less than the outermost radius $R$). Note that the shear strain developed on the outermost surface of the specimen (at radius $R$) will be the maximum value ($\gamma_{max}$) of shear strain in the specimen.

The value of shear stress at the outermost surface of the specimen obtained from the maximum value of shear strain will also result in the maximum value of shear stress. The relationship between the maximum shear stress and angle of twist can be obtained by combining Eqs. (4.15) and (4.16):

$$\tau_{max} = GR\frac{\varnothing}{L}. \tag{4.17}$$

The value of maximum shear stress can also be obtained in terms of applied torque using the torsion formula:

$$\tau_{max} = \frac{TR}{I_p}, \tag{4.18}$$

where $T$ is the value of applied torque and $I_p$ is the polar moment of inertia of the circular cross section. For a circular cross section of radius $R$, the polar moment of inertia can be defined as:

$$I_P = \frac{\pi R^4}{2}. \tag{4.19}$$

Mechanical properties such as shear modulus ($G$), ultimate shear strength ($\tau_u$), and yield shear strength ($\tau_y$) can be determined from the shear stress vs. shear strain or torque vs. angle of twist behavior of the specimen similar to the method described in Sect. 4.2.2.

**Example 4.3**
A circular dumbbell-shaped specimen (with a gauge diameter of 3 mm) of human tibia was subjected to torsion. The torque applied to the specimen up to the elastic limit is 25 N-mm and the torsion modulus of elasticity of the bone specimen was recorded as 3.2 GPa. Calculate the (a) maximum shear stress developed in the specimen and (b) angle of twist of the specimen per unit length.

**Solution**
Given: Diameter of the specimen $d = 3$ mm, radius $R = 1.5$ mm, applied torque $T = 25$ N-mm, and torsional modulus $G = 3.2$ GPa $= 3200$ MPa.

(a) The maximum shear stress developed in the specimen can be calculated using $\tau_{\max} = \frac{TR}{I_p}$. The polar moment of inertia ($I_p$) in this equation can be determined as follows:

$$I_P = \frac{\pi R^4}{2} = \frac{3.14 \times 1.5^4}{2} = 7.95 \; mm^4.$$

Therefore, $\tau_{\max} = \frac{TR}{I_p} = \frac{25 \times 1.5}{7.95} = \mathbf{4.72 \, MPa}$.

(b) The angle of twist per unit length can be calculated using the relationship $\tau_{\max} = GR\frac{\varnothing}{L}$. Therefore, the angle of twist $\frac{\varnothing}{L} = \frac{\tau_{\max}}{GR} = \frac{4.72}{3200 \times 1.5} = \mathbf{9.83 \times 10^{-4} \, radians/mm}$.

## 4.3  Viscoelastic Properties and Dynamic Testing

### 4.3.1  Viscoelastic Properties

Viscoelasticity is a dynamic behavior of specimens that exhibit both viscous and elastic characteristics when undergoing deformation. Viscous materials (like water) resist shear flow if a shear stress is applied, while elastic materials (like metals) deform under the force or stress and are able to return to their original state upon removal of the stress.

Many native tissues, crosslinked biomaterials (such as hydrogels), and biomaterial scaffolds exhibit viscoelastic properties. In native tissues, such as muscle, bone, cartilage, and tendon, the viscoelastic properties play critical roles in regulating development, physiology, and pathophysiology via the extracellular matrix (ECM). Biomaterials and/or printed scaffolds are artificial ECM, and their viscoelasticities have substantial impacts on cell functioning, including attachment, proliferation, and differentiation [1]. In both native tissues and biomaterial scaffolds, the ECM transmits external mechanical forces to the cells within, thus modulating cell behavior and guiding cell fate. Such interactions of cells with the ECM occur through a complex process in which viscoelasticity plays an important role.

Notably, the mechanical properties discussed in the previous sections are known as static mechanical properties and are measured by means of static testing or applying loading or displacement with a slow rate of change. In contrast, viscoelastic properties are dynamic properties measured by means of dynamic testing where the rates of loading or displacement affect the testing results. Dynamic testing methods typically include cyclic, creep, stress-relaxation, and frequency-dependent testing, discussed as follows.

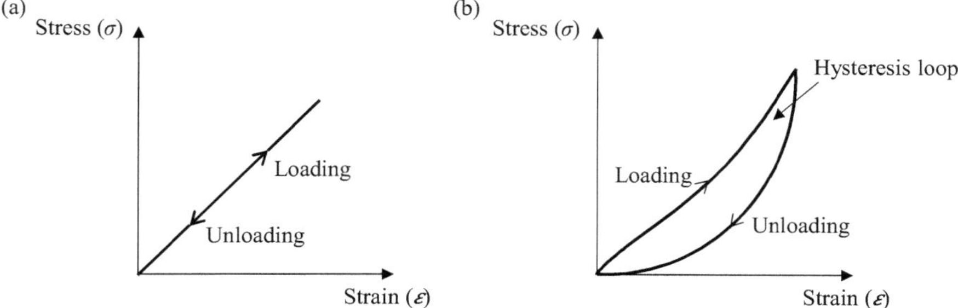


**Fig. 4.12** Stress-strain curves of a (**a**) purely elastic material and (**b**) viscoelastic material observed in a cyclic loading test

## *4.3.2 Cyclic Load Tests*

In a *cyclic loading test*, the loading and unloading of stress or strain (typically by compressive or tensile force or displacement) is cyclically applied to the sample at a defined rate. Figure 4.12 shows the stress-strain curves of both a purely elastic material and viscoelastic material. When the stress is applied to a purely elastic material, the material deforms linearly with the applied stress and returns to its original shape upon removal of the stress. In contrast, a viscoelastic material deforms with the applied stress both elastically and plastically (or by shear flow); such behavior is characterized by a hysteresis loop in the stress-strain curve (Fig. 4.12b). The area inside the hysteresis loop is associated with the energy lost (as heat) during the loading cycle, thus indicating the degree of viscosity in the viscoelastic sample. The hysteresis is dependent on both the rate and magnitude of stress or strain applied in testing.

## *4.3.3 Creep and Stress-Relaxation Tests*

The creep and stress-relaxation properties of a viscoelastic material are determined by performing tests under constant stress and strain loading, respectively. In a *creep test*, a constant load or stress (typically compressive or tensile stress) is applied to the sample while the sample deformation or strain of the sample is recorded over time. A typical strain time plot for such a test is illustrated in Fig. 4.13. Initially, the sample reacts to the load with an instantaneous response and reaches a certain value of elastic strain, $\varepsilon_0$. As the load continues to be applied, the sample experiences additional plastic deformation, which is characterized by an increase in strain with time, but at a decreased rate (or slope of the curve) due to dislocations in the material in response to the applied load. To quantitatively characterize the creep response, a parameter for creep ($\tau_{3/2}$) is commonly used and defined by the time for the strain to

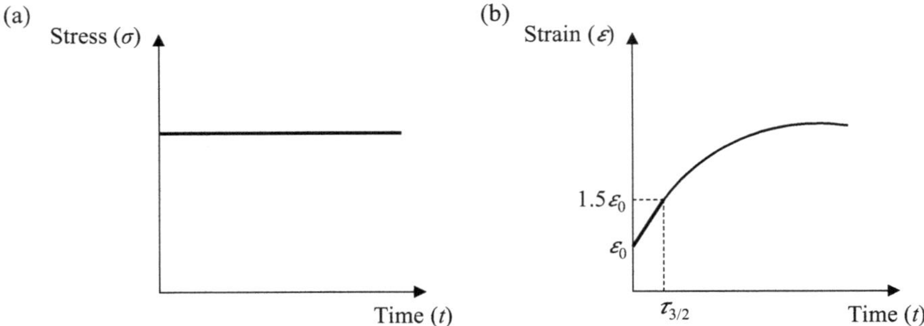

**Fig. 4.13** Creep testing: (**a**) constant stress applied on the specimen and (**b**) creep response, starting with an instantaneous elastic deformation followed by plastic deformation with continually increasing strain

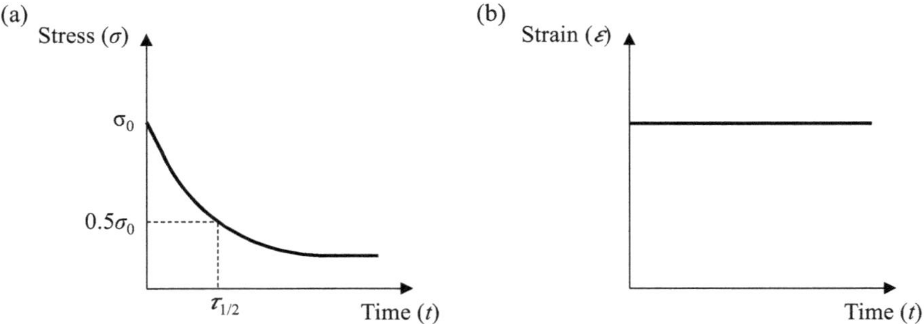

**Fig. 4.14** Stress-relaxation testing: (**a**) constant strain applied on the specimen and (**b**) relaxation response, starting with an instantaneous elastic deformation followed by a diminishing plastic deformation with continually increasing strain

reach 1.5 times or 150% of the initial elastic strain. Both stress magnitude and temperature may affect the creep response, including the value of $\tau_{3/2}$.

In a *stress-relaxation test*, the sample is loaded with a constant deformation or strain while the stress induced in the sample is recorded over time. The stress-relaxation response, as illustrated in Fig. 4.14, starts with an instantaneous elastic stress, $\sigma_0$, followed by the stress diminishing with time. To characterize the relaxation response, a half-stress relaxation time ($\tau_{1/2}$) is usually used and is defined by the time for the stress to relax and reach 0.5 times or 50% of the initial elastic stress. Both strain magnitude and temperature may affect the relaxation, resulting in different values of $\tau_{1/2}$.

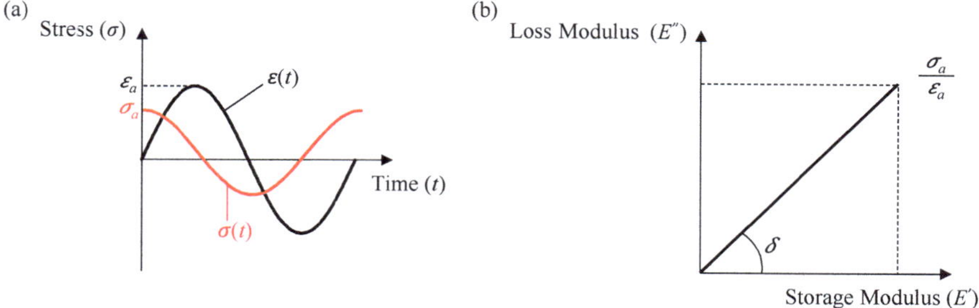



**Fig. 4.15** Frequency-dependent testing: (**a**) sinusoidal stress and strain and (**b**) illustration of storage modulus, loss modulus, and loss angle

### *4.3.4   Frequency-Dependent Tests*

In a *frequency-dependent test*, a sinusoid stress/strain is imposed on the samples with the resulting strain/stress recorded. The stress/strain can be generated by applying tensile or compressive force/displacement on extensional rheometers (known as extensiometers) discussed below or by shear force/displacement on shear rheometers as discussed in Chap. 5.

Stress and strain plots in a tensile or compressive loading test are shown in Fig. 4.15a, with the strain and stress mathematically given by, respectively,

$$\varepsilon(t) = \varepsilon_a \sin(\omega t), \tag{4.20}$$

and

$$\sigma(t) = \sigma_a \sin(\omega t + \delta), \tag{4.21}$$

where $\omega$ is the frequency or angular velocity of the strain oscillation, $t$ is time, $\delta$ is the phase lag between stress and strain (or loss angle), and $\varepsilon_a$ and $\sigma_a$ are the amplitudes of strain and stress, respectively.

The complex modulus $E(\omega)$ is the ratio of stress to strain under the sinusoidal loading for a viscoelastic material and is given by:

$$E(\omega) = \frac{\sigma(\omega)}{\varepsilon(\omega)} = \frac{\sigma_a e^{i(\omega t + \delta)}}{\varepsilon_a e^{i\omega t}} = \frac{\sigma_a}{\varepsilon_a} e^{i\delta} = \frac{\sigma_a}{\varepsilon_a} \cos(\delta) + i \frac{\sigma_a}{\varepsilon_a} \sin(\delta) = E' + iE''. \tag{4.22}$$

The complex modulus $E(\omega)$ consists of two components that are respectively associated with the amount of energy stored in the material and the energy dissipated as heat. The stored energy is related to the elastic behavior of the material and is represented by the storage modulus:

$$E' = \frac{\sigma_a}{\varepsilon_a} \cos(\delta). \tag{4.23}$$

The loss of energy due to heat dissipation from the viscous behavior of the material is represented by the loss modulus:

$$E'' = \frac{\sigma_a}{\varepsilon_a} \sin(\delta). \tag{4.24}$$

The relationships among the storage modulus, loss modulus, and loss angle are illustrated in Fig. 4.15b, where $\tan(\delta) = E''/E'$. For a given material or solution, the storage modulus, loss modulus, and loss angle are dependent on the frequency ($\omega$) and amplitude ($\varepsilon_a$) of the sinusoidal strain applied, which also explains why such testing is called frequency-dependent testing.

### 4.3.5 Mathematical Models of Linear Viscoelastic Behavior

Linear viscoelastic behavior has elastic and viscous components that can be modeled using a combination of spring and dashpot elements. In these viscoelastic models, the springs represent the elastic response and the dashpots characterize the viscous response. A spring behaves according to Hook's law:

$$\sigma = E\varepsilon, \tag{4.25}$$

which in differential form is:

$$\frac{d\sigma}{dt} = E\frac{d\varepsilon}{dt}. \tag{4.26}$$

A dashpot, like a shock absorber in a car, follows Newton's law:

$$\sigma = \eta\frac{d\varepsilon}{dt}. \tag{4.27}$$

In the above equations, $E$ is the elastic modulus of the spring component and $\eta$ is the viscosity of the dashpot element. There are varying ways to combine the springs and dashpots to represent viscoelastic behavior. This results in different models, among which the Maxwell and Kelvin-Voigt models are the most common.

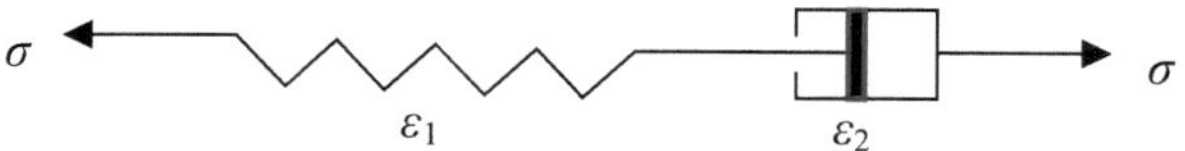


**Fig. 4.16** Schematic of the Maxwell model consisting of a spring and a dashpot in series

### 4.3.5.1   Maxwell Model

The Maxwell model consists of a spring and a dashpot in series, as shown in Fig. 4.16. When a stress $\sigma$ is applied, two strains $\varepsilon_1$ and $\varepsilon_2$ are produced on the spring and dashpot elements, respectively. Thus, the total strain $\varepsilon$ produced in this model is the sum of the individual strains as given by:

$$\varepsilon = \varepsilon_1 + \varepsilon_2. \tag{4.28}$$

Given that the spring and dashpot elements are in series, the stress on each element is the same as the applied stress $\sigma$.

The differential form of Eq. (4.28) is:

$$\frac{d\varepsilon}{dt} = \frac{d\varepsilon_1}{dt} + \frac{d\varepsilon_2}{dt}. \tag{4.29}$$

This equation can be rearranged in terms of the applied stress by applying the laws given in Eqs. (4.26) and (4.27) for the spring and dashpot elements, respectively, as given by:

$$\frac{d\sigma}{dt} = E\frac{d\varepsilon_1}{dt}, \tag{4.30}$$

and

$$\sigma = \eta\frac{d\varepsilon_2}{dt}, \tag{4.31}$$

resulting in the strain given by:

$$\frac{d\varepsilon}{dt} = \frac{1}{E}\frac{d\sigma}{dt} + \frac{\sigma}{\eta}. \tag{4.32}$$

Equation (4.32) can be used to represent the creep and stress relaxation behavior of viscoelastic materials. For the creep test, a constant stress ($\sigma_o$) is applied to examine the resulting strain over a given time period. As such, the rate of change of applied stress with time is zero, that is, $d\sigma/dt = 0$. Equation (4.32) in this case is reduced to:

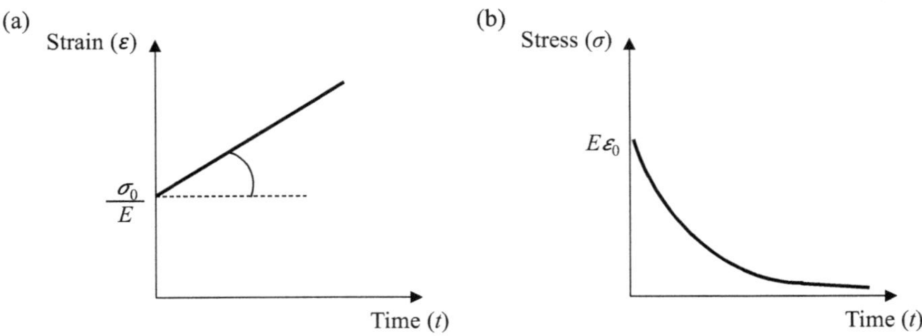

**Fig. 4.17** Responses of (**a**) creep and (**b**) stress relaxation, as predicted by the Maxwell model

$$\frac{d\varepsilon}{dt} = \frac{\sigma_o}{\eta}. \tag{4.33}$$

This resulting equation shows the strain rate is constant over time. In other words, the value of strain increases linearly as shown in Fig. 4.17a, where the strain has an initial value of $\frac{\sigma_o}{E}$, determined by Eq. (4.25). Note that viscoelastic materials rarely follow the creep behavior shown in Fig. 4.17a and thus this model is generally considered ineffective for predicting the creep behavior of a viscoelastic material.

For the stress-relaxation test, the value of strain is kept constant ($\varepsilon_o$). Thus, the rate of change of strain with time is zero, that is, $d\varepsilon/dt = 0$, and Eq. (4.32) becomes:

$$0 = \frac{1}{E}\frac{d\sigma}{dt} + \frac{\sigma}{\eta}. \tag{4.34}$$

Rearranging this equation yields:

$$\frac{d\sigma}{\sigma} = -\frac{E}{\eta}dt. \tag{4.35}$$

Integrating this resulting equation yields the stress expressed as a function of time, that is,

$$\log{_e}\sigma = -\frac{E}{\eta}t + C, \tag{4.36}$$

where $C$ is an integral constant given by, when $t = 0$,

$$C = \log{_e}\sigma_o. \tag{4.37}$$

Note that the stress ($\sigma_o$) is the stress caused by the applied strain $\varepsilon_o$ at the beginning of the test and given by $E\varepsilon_o$ from Eq. (4.25). Thus, Eq. (4.36) can be rewritten as:

$$\log {}_e\sigma = -\frac{E}{\eta}t + \log {}_e\sigma_o. \tag{4.38}$$

or

$$\log {}_e\sigma - \log {}_e\sigma_0 = -\frac{E}{\eta}t. \tag{4.39}$$

Applying the logarithmic rules to Eq. (4.39) can yield:

$$\sigma = \sigma_o e^{-\frac{E}{\eta}t}. \tag{4.40}$$

This equation indicates the value of stress during a stress relaxation test decreases exponentially, as shown in Fig. 4.17b. This type of behavior is observed for many biomaterials in tissue engineering.

### 4.3.5.2  Kelvin-Voigt Model

In the Kelvin-Voigt model, the spring and dashpot elements are connected in parallel, as shown in Fig. 4.18. In this case, the applied load or stress is split to apply to both elements, generating the same deformation or strain. Thus, one has:

$$\sigma = \sigma_1 + \sigma_2 \tag{4.41}$$

and

$$\varepsilon = \varepsilon_1 = \varepsilon_2. \tag{4.42}$$

From Eqs. (4.25) and (4.27), the stress generated on the spring and dashpot elements can be determined by:

$$\sigma_1 = E\varepsilon_1. \tag{4.43}$$

and

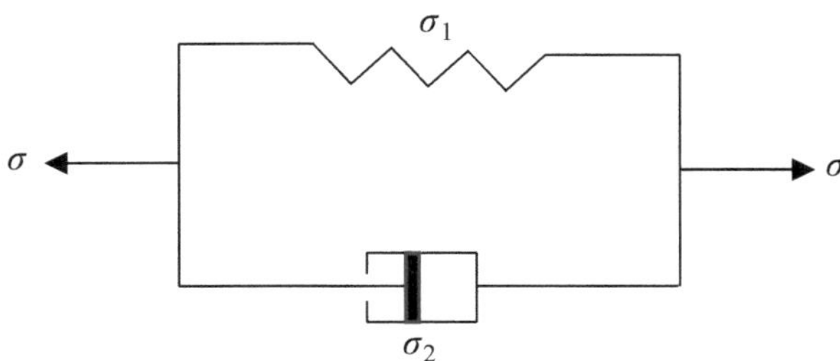


**Fig. 4.18** Schematic of the Kelvin-Voigt model consisting of a spring and a dashpot in parallel

$$\sigma_2 = \eta \frac{d\varepsilon_2}{dt}. \tag{4.44}$$

Substituting Eqs. (4.42), (4.43), and (4.44) into Eq. (4.41) can yield:

$$\sigma = E\varepsilon + \eta \frac{d\varepsilon}{dt}. \tag{4.45}$$

From the above equation, the change rate of strain is given by:

$$\frac{d\varepsilon}{dt} = \frac{\sigma}{\eta} - \frac{E\varepsilon}{\eta}. \tag{4.46}$$

For the creep test where the applied stress remains constant, that is, $\sigma = \sigma_o$, Eq. (4.46) becomes:

$$\frac{d\varepsilon}{dt} = \frac{\sigma_o}{\eta} - \frac{E\varepsilon}{\eta}. \tag{4.47}$$

Solving the above equation yields:

$$\varepsilon = \frac{\sigma_o}{E} \left( 1 - e^{\frac{-E}{\eta}t} \right). \tag{4.48}$$

Equation (4.48) shows that, for the creep test, the Kelvin-Voigt model predicts strain as increasing exponentially with time; as time approaches infinity, the value of strain becomes a constant given by $\sigma_o/E$. The strain vs. time plot for this model is illustrated in Fig. 4.19.

For the stress relaxation test where the value of strain is kept constant at $\varepsilon_0$ or $d\varepsilon/dt = 0$, Eq. (4.47) becomes:

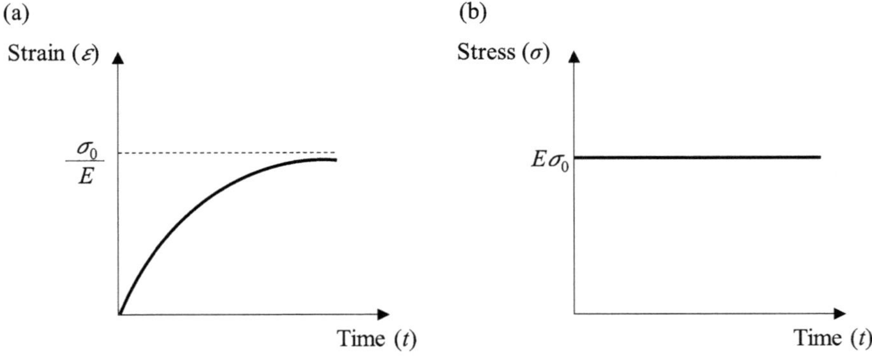


**Fig. 4.19** Responses of (**a**) creep and (**b**) stress relaxation, as predicted by the Kelvin-Voigt model

$$0 = \frac{\sigma}{\eta} - \frac{E\varepsilon_0}{\eta} \tag{4.49}$$

or

$$\sigma = E\varepsilon_0. \tag{4.50}$$

The above equation shows the Kelvin-Voigt model for the stress relaxation test does not involve the contribution of the viscous dashpot and only gives the response of the spring. As such, the Kelvin-Voigt model is not considered effective for stress relaxation testing.

**Example 4.4**

An alginate sample was produced from a 5% w/v alginate solution using a cylindrical mold and a 100 mM $CaCl_2$ cross-linking solution. The sample has an area of 52 mm$^2$ and a height of 4.6 mm. The sample was subjected to stress-relaxation testing by applying a step displacement of 1.5 mm and recording the stress relaxation over a 27 s time period on a dynamic testing machine. The resulting data are given in the table below. Fit the measured stress relaxation behavior of this sample using the Maxwell model and determine the elastic and viscosity constants of the model, and then estimate the half-stress relaxation time ($\tau_{1/2}$).

| Time (s) | Load (N) |
| --- | --- |
| 0 | 4.220 |
| 0.018 | 3.990 |
| 0.022 | 3.787 |
| 0.102 | 3.536 |
| 0.248 | 3.354 |
| 0.621 | 3.177 |
| 1.118 | 2.958 |
| 1.988 | 2.739 |
| 3.11 | 2.593 |
| 5.093 | 2.458 |
| 7.577 | 2.354 |
| 9.441 | 2.271 |
| 11.180 | 2.218 |
| 14.037 | 2.1465 |
| 16.646 | 2.104 |
| 19.503 | 2.052 |
| 22.236 | 2.010 |
| 24.590 | 1.979 |
| 26.956 | 1.969 |

**Solution**

At time $t = 0$, the constant strain value $\varepsilon_o$ = displacement/ height = 1.5/4.6 = 0.33, while the initial stress $\sigma_o$ = load/area = 4.22/52 = 0.0812 MPa = 81.2 kPa. Thus, the elastic constant $E = \sigma_0/\varepsilon_0$ = 81.2/0.33 = 246.1 kPa.

For the stress relaxation, the Maxwell model is:

$$\sigma = \sigma_o e^{-\frac{E}{\eta} t}$$

or

$$\frac{\sigma}{\sigma_o} = e^{-\frac{E}{\eta} t}.$$

As such, the data listed in the table can be fitted using the above exponential equation, where the values of stress, $\sigma$, can be obtained by dividing load values in the table by the sample area. The stress values can be then converted from MPa to kPa.

The $E/\eta$ value resulting from this calculation is 0.046. The elastic constant, $E$, is already determined above to have a value of 246.1 kPa. Therefore, the value of viscous constant $\eta$ = 246.1/0.046 = 5350 kPa.s. The Maxwell model for the stress relaxation can therefore be written as

$$\sigma = 81.2 \; e^{\frac{-246.1}{5350} t}.$$

The half-stress relaxation time $(\tau_{1/2})$ is the time for the stress to relax and reach 0.5 times or 50% of the initial elastic stress, i.e., $\sigma_o/2$ = 81.2 kPa/2 = 40.6 kPa. Substituting it for $\sigma$ in the above Maxwell model results in a $\tau_{1/2}$ value of 15.07 s.


## 4.4   Mechanical-Property Measurements of Native Tissues and Scaffolds

### 4.4.1   Sample Preparation

An important step before performing any desired mechanical testing is preparing the specimen. Preparation of biological specimens is a challenging task as native tissues are typically anisotropic, heterogeneous, and viscoelastic, which are factors that must be considered in addition to complications resulting from limited tissue availability. Tissue samples should be extracted from the desired anatomic location and tested along specific directions (e.g., the diaphysis of a cortical bone) at appropriate strain/ loading rates to avoid material complications associated with anisotropy, heterogeneity, and viscoelasticity. During sample preparation, the sectioning process on machines or tools to cut or section samples from the main source may cause heat that can alter the mechanical properties of the samples obtained. As such, samples should be kept cool (i.e., with lubricant and/or coolant) during sectioning.

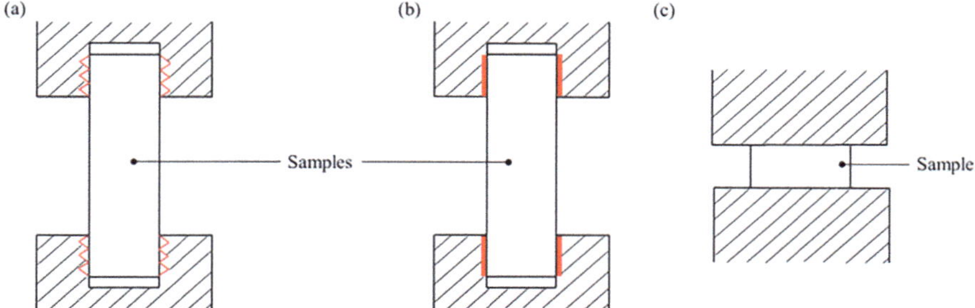

**Fig. 4.20** Sampling mounting methods: (**a**) mechanical clamping, (**b**) adhesive clamping, and (**c**) placement in the space between two plates of the testing apparatus for compressive loading

## 4.4.2  Considerations During the Testing

During mechanical testing, a number of testing conditions can affect the measured results or mechanical properties of native tissue and scaffolds and therefore must be considered.

**Sample Mounting**  Samples must be mounted onto the mechanical testing instruments as appropriate for loading. In tensile testing, the sample is typically mounted by employing either a mechanical or adhesive clamp, as shown in Fig. 4.20a, b, respectively. Mechanical clamping is more suitable for samples with strong mechanical properties, while adhesive clamping is preferred for delicate samples with weak mechanical properties. Important here is that a proper clamping method should be selected and utilized to avoid non-uniform gripping and damage to the sample before testing. For compressive testing, the sample is mounted between two plates that can be controlled to move relative to one another, as shown in Fig. 4.20c.

**Operating Temperature and Humidity**  The mechanical behavior of a sample can be affected by humidity and temperature, so maintaining these parameters at appropriate and controlled levels is important. For example, samples may be installed or placed in the test chamber on a testing instrument such as the one in Fig. 4.1, where a media of, for example, phosphate buffered saline (PBS), can be circulated throughout to control or adjust the temperature and humidity.

**Loading Direction**  For isotropic samples, mechanical properties are independent of loading direction. In other words, the samples may be oriented in any manner during testing without affecting the measured data. However, almost all tissues and scaffolds display anisotropy, or the dependence of mechanical properties on the loading direction. For example, three-dimensional printed tissue scaffolds display either transversely isotropic or orthotropic tensile and compressive properties. Testing such scaffolds in multiple orientations to generate unique stress-strain curves for each different loading direction is therefore required. For transversely isotropic scaffolds, properties must be tested in both the longitudinal and transverse

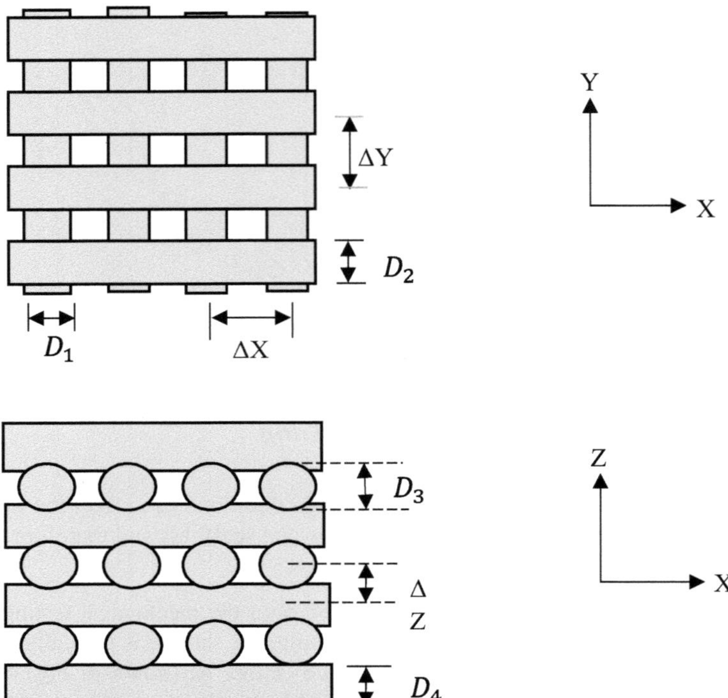

**Fig. 4.21** Microstructures of transversely isotropic and orthotropic rapid prototyping fabricated tissue scaffolds

directions. For orthotropic scaffolds, three separate tests must be conducted to determine the mechanical properties of interest in all three directions corresponding to the normal axes of a Cartesian coordinate system. Figure 4.21 illustrates an example of transversely isotropic and orthotropic microstructures. For transverse isotropy, $\Delta X = \Delta Y$, $D_1 = D_2$, and $D_3 = D_4$. For orthotropic anisotropy, at least one of the following must be true: $\Delta X \neq \Delta Y$, $D_1 \neq D_2$, or $D_3 \neq D_4$.

### 4.4.3  Case Studies: Measurement of Mechanical Properties

**Case Study 4.1**

*Measurement of Mechanical Properties of Hard Tissue (Cortical Bone).* Bone is considered a highly heterogeneous, anisotropic, and hierarchical material. At the lowest level of its microstructure, bone consists of polymeric strands of tropocollagen molecules crystallized with apatite at the molecule strand ends. These molecules then form collagen fibrils that are themselves bundled into individual collagen fibers.

These fibers are arranged in a plywood-like microstructure to form lamella layers that are concentrically organized into tube-like structures known as osteons. Bone itself consists of an outer shell of dense tissue known as compact or cortical bone, an inner core of porous cellular material called cancellous or trabecular bone, and an outer membrane known as the periosteum. When injury/disease to bone tissue exceeds its ability to self-repair, bone tissue must be harvested to fill the injury and allow the wound to be bridged by regenerating tissue. The amount of tissue that can be harvested from a patient is limited and donor tissue can lead to an immune system response and tissue rejection. To improve the treatment of these types of injuries, bioprinted scaffolds can be employed to for bone reconstitution and treatment [2]. Tissue scaffolds should match the properties of the tissue they replace, and ideally maintain these properties throughout the entire treatment time.

This case study focuses on the measurement of tensile properties of cortical bone using a uniaxial tensile test, based a previous study [3]. The tensile test presented in this section was conducted on femoral cortical bone obtained from a young goat (about 30 months of age). The mechanical properties of fresh bone tissue can abruptly change if it is allowed to dry before testing. The following methods were adopted for optimal preservation of the mechanical properties of cortical bone. After removing the femoral bone from the animal and cleaning it of surrounding soft tissue, it was wrapped in gauze soaked in normal saline. The bone was further wrapped in plastic wrap and sealed in an airtight plastic bag. The plastic bag was stored in a freezer at $-20$ °C within 1 h of harvest. The femoral bone was kept hydrated in saline solution after removal from the freezer and during all stages of tissue preparation. The tensile mechanical properties of the bone were determined using the following steps.

(a) *Specimen Preparation for Tensile Testing.*

The cortical bone specimen for the tensile test was prepared in a dumbbell shape with a thickness of 2.5 mm, gauge length of 25 mm, gauge width of 4 mm, and total length of 80 mm for the test in the longitudinal direction (load applied along the long axis of the bone). To prepare the bone specimen, rough cuts were made using a hacksaw blade and the overheated surrounding area was removed with wet sandpaper. Fine cuts were made using a diamond cutter (Isomet 4000) and wire hacksaw. Flattening of the specimen was carried out using a belt sander and fillets were made using needle files. The bone specimen was irrigated with water and saline during the different cutting and machining processes to prevent heating. The bone specimen was preserved at room temperature in a solution of 50% saline and 50% ethanol until testing. The dumbbell-shaped cortical bone specimen prepared for tensile testing is shown in Fig. 4.22b, along with its mounting on the test machine (Fig. 4.22a).

(b) *Testing Procedure.*

The ends of the dumbbell specimen were clamped in the upper and lower jaws of the machine, and an extensometer with a gauge length of 25 mm was attached to the gauge region to measure the displacement during tensile loading. The tensile load during this test was applied using a 5-kN load cell. The bone

**Fig. 4.22** Tensile test performed on cortical bone specimen: (**a**) bone specimen clamped on the testing machine, (**b**) dumbbell specimen prepared from cortical bone, and (**c**) fractured bone specimen obtained after tensile test

specimen was loaded to the fracture point and results obtained in the form of load-displacement data points. The broken specimen of cortical bone after the tensile test is shown in Fig. 4.22c.

(c) *Results of Tensile Test Performed on Cortical Bone Specimen.*

The results of the tensile test were obtained in the form of engineering stress and strain data points as reported in Table 4.1. These data points can be used to plot the stress-strain behavior of the cortical bone and to evaluate its tensile properties in terms of the elastic modulus, yield strength, and ultimate strength. The yield strength can be measured using the 0.2% offset method, and the elastic modulus can be determined using stress and strain values from the initial linear portion of the stress-strain curve as discussed in Example 1 in Sect. 4.2.2. The stress-strain curve drawn using the data points reported in Table 4.1 is shown in Fig. 4.23.

1. Three sets of data are selected from the linear portion of the stress-strain curve to determine the average value of the elastic modulus ($E$), that is,

$$E_1 = \frac{20.08 - 12.05}{0.0008 - 0.0005} = 26766.7 \, MPa$$

**Table 4.1** Values of engineering stress and strain obtained from tensile testing

| Strain (mm/mm) | Stress (MPa) |
| --- | --- |
| 0.0000 | 0.00 |
| 0.0005 | 12.05 |
| 0.0008 | 20.08 |
| 0.0011 | 28.11 |
| 0.0014 | 36.14 |
| 0.0019 | 48.19 |
| 0.0025 | 64.25 |
| 0.0030 | 76.30 |
| 0.0033 | 84.33 |
| 0.0038 | 95.95 |
| 0.0041 | 104.73 |
| 0.0045 | 111.16 |
| 0.0049 | 117.04 |
| 0.0062 | 125.00 |
| 0.0083 | 131.08 |
| 0.0110 | 134.46 |
| 0.0132 | 136.49 |
| 0.0158 | 137.84 |
| 0.0178 | 138.51 |
| 0.0204 | 139.87 |
| 0.0233 | 141.22 |
| 0.0261 | 142.57 |
| 0.0298 | 143.24 |
| 0.0333 | 144.60 |
| 0.0354 | 144.60 |

$$E_2 = \frac{28.11 - 20.08}{0.0011 - 0.0008} = 26766.7 \, MPa$$

$$E_3 = \frac{36.14 - 28.11}{0.0014 - 0.0011} = 26766.7 \, MPa$$

Therefore, $E$ is the average of $E_1$, $E_2$, and $E_3$ and is equal to 26766.7 MPa or 26.67 GPa.

2. The yield strength calculated using the 0.2% offset method as shown in Fig. 4.23 is 128.3 MPa.
3. The ultimate strength is equal to the maximum stress value, i.e., 144.60 MPa from Table 4.1.

**Case Study 4.2**
*Measurement of Mechanical Properties of Printed Scaffolds.* The properties of printed scaffolds in terms of compressive Young's modulus and compressive

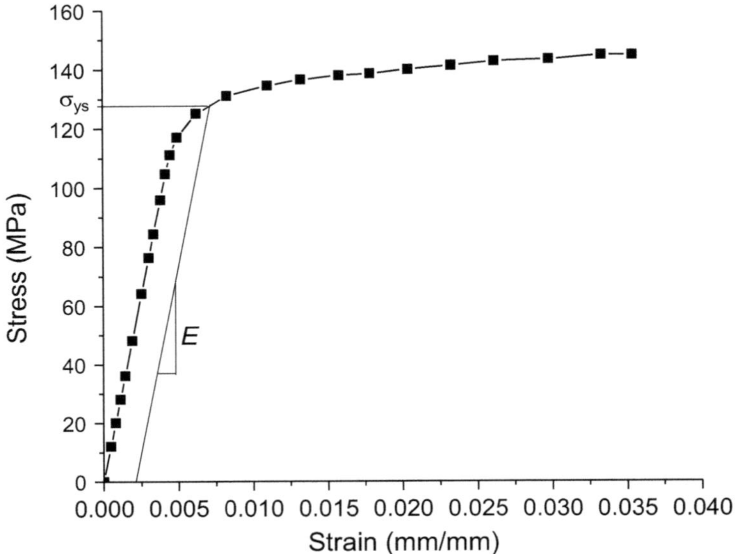

**Fig. 4.23** Stress-strain curve obtained for femoral cortical bone under tension

strength are important in tissue engineering and modeling and are examined in this case study based on previous research [4]. The compression test is performed on a universal testing machine (UTM) using the appropriate load cell according to the strength of the tested material. Specimens of tissue scaffolds for compression tests are prepared with either a cuboid or cylindrical shape. The load is applied in a static manner at a very small displacement rate of about 1 mm/min. The results are obtained in terms of load-displacement curves. The load-displacement curves can be converted into stress-strain curves using different mechanical relationships as discussed in the previous sections.

(a) *Scaffold Design and Fabrication for Compression Testing.*

The compressive behavior of a 3D scaffold made of poly($\varepsilon$)-caprolactone (PCL) (molecular weight 80,000 g·mol$^{-1}$) is examined and discussed in this case study. The computer-aided design (CAD) model of the PCL scaffold was designed in Magics Envisontec. This CAD model was exported into Bioplotter RP and sliced into 10 horizontal layers. The scaffold was fabricated using a cylindrical metal needle with an inner diameter of 300 μm at a crosshead speed of 1 mm/s, pressure of 0.8 MPa, and temperature of 110 °C. This produced a cuboid geometry porous scaffold with dimensions of 15 × 15 × 2.4 mm$^3$, 10 layers, a strand thickness of 0.26 mm, and a 1-mm interstrand spacing with a 0–90° pattern.

(b) *Testing Procedure.*

The uniaxial compression test was conducted on an Instron 3366 material testing machine. The scaffold was placed between two smooth and rigid platens of the Instron machine after measuring its gauge length, width, and thickness. Mechanical load was applied using a 1 kN load cell at a crosshead speed of 1 mm/min. The load-displacement behavior of the scaffold was recorded throughout the experimental testing, with stress-strain values evaluated using the expressions discussed in the previous sections to plot the stress-strain curve of the scaffold under compression loading.

(c) *Experimental Results.*

The stress and strain values obtained during the compression test are reported in Table 4.2. The corresponding stress-strain behavior is shown in Fig. 4.24, with the stress-strain relationship demonstrating a bilinear nature. The elastic modulus of the scaffold was determined from the initial slope of the stress-strain curve, which can be easily determined by linear regression of the initial data points as shown in Fig. 4.24.

The slope of the initial part of the stress-strain curve obtained from the linear regression shown in Fig. 4.25 was found to be 40.54 MPa with an $R^2$ value of 0.96. This means that 96% of the data points fit the linear behavior and the elastic modulus of the PCL scaffold is about 40.54 MPa. The compressive strength of this scaffold

**Table 4.2** Engineering stress and strain values obtained from compression testing

| Strain (mm/mm) | Stress (MPa) |
| --- | --- |
| 0.00 | 0.00 |
| 0.06 | 0.83 |
| 0.13 | 2.74 |
| 0.18 | 4.38 |
| 0.24 | 6.30 |
| 0.29 | 8.21 |
| 0.34 | 10.39 |
| 0.40 | 14.21 |
| 0.46 | 17.48 |
| 0.50 | 20.21 |
| 0.55 | 24.02 |
| 0.64 | 30.29 |
| 0.72 | 36.83 |
| 0.82 | 45.55 |
| 0.91 | 54.26 |
| 1.01 | 63.25 |
| 1.09 | 71.96 |
| 1.20 | 82.86 |
| 1.29 | 92.66 |
| 1.37 | 101.65 |
| 1.49 | 117.16 |

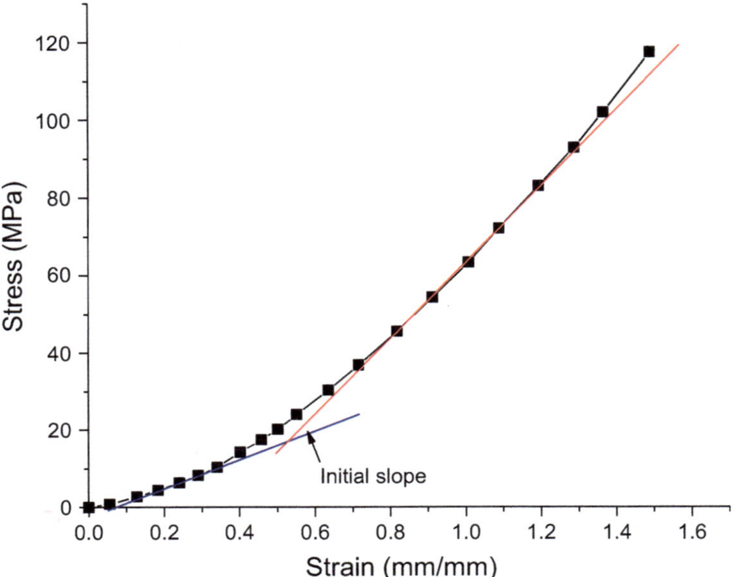



**Fig. 4.24** Stress and strain behavior of porous PCL scaffold under compression loading

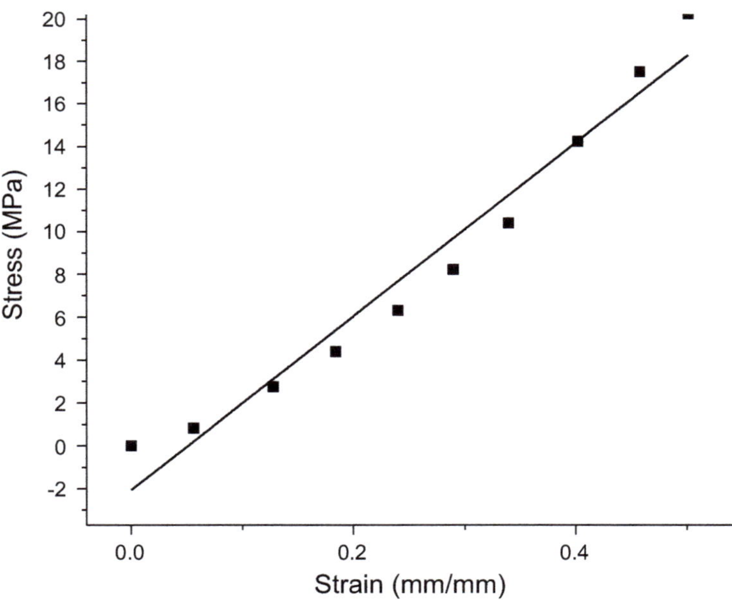



**Fig. 4.25** Linear curve fit of initial stress-strain data points to evaluate the elastic modulus

corresponding to the maximum value of compressive stress was found to be 117.16 MPa, as reported in Table 4.2.

**Case Study 2:**

Viscoelastic Behavior of Alginate Hydrogel. Alginate hydrogels have various applications in the area of tissue engineering due to their low toxicity, biocompatibility, and mild gelation. These hydrogels have a structure similar to the ECM of different living tissues that therefore allows for wide applications in wound healing and cell transportation. This case study was conducted to examine and understand the viscoelastic behavior of alginate specimens prepared from 4% and 5% alginate solutions after crosslinking with 100 mM calcium chloride solution [5]. The viscoelastic behavior was analyzed under both static and dynamic loading conditions using a dynamic testing machine.

(a) *Sample Preparation for Mechanical Testing.*

Two different 25 mL solutions with 4% and 5% w/v alginate were prepared by, respectively, adding 1.0 and 1.25 g of alginate to a beaker and then pouring distilled water into the beaker up to the 25 mL mark. These solutions were mixed for more then 3 h using a magnetic stirring machine at room temperature. The alginate solutions were poured in cylindrical molds each placed inside a beaker and then 100 mM $CaCl_2$ crosslinking solution was carefully poured into each beaker. The beakers were then covered and refrigerated for about 24 h to allow crosslinking of the alginate solution.

After crosslinking, the alginate solution was converted into a solid gel-like material in the cylindrical molds. The resulting 4% and 5% alginate cylinders were then cut into pieces to make specimens with the desired dimensions (height $4.6 \pm 0.6$ mm and area $52.97 \pm 5.67$ mm$^2$ for 4% alginate; height $3.2 \pm 0.3$ mm and area $52.36 \pm 4.33$ mm$^2$ for 5% alginate) for mechanical testing.

(b) *Testing Procedure.*

Static and dynamic tests were performed under displacement control on a dynamic testing machine (ElectroForce BioDynmic™, Bose, USA)) with a load capacity of 20 N. Static compression tests were performed at two displacement rates (0.005 and 0.01 mm/s) for both the 4% and 5% alginate hydrogels. Dynamic tests were conducted using the sinusoidal wave option (0.5 Hz) for the displacement rate and frequency. The load-displacement curves for both the static and dynamic tests were recorded and converted into stress-strain curves using the equations discussed in previous sections.

(c) *Experimental Results.*

The stress-strain curves obtained for the 4% and 5% alginate materials for two different displacement rates under static compressive loading are shown in Fig. 4.26.

The stress-strain behavior of the alginate materials was found to be similar to that of soft biological tissue. The stress-strain curves consist of an initial curved zone with some radius of curvature followed by a linear region, as shown in Fig. 4.26. This is typical behavior for a viscoelastic material; the curved zone represents the

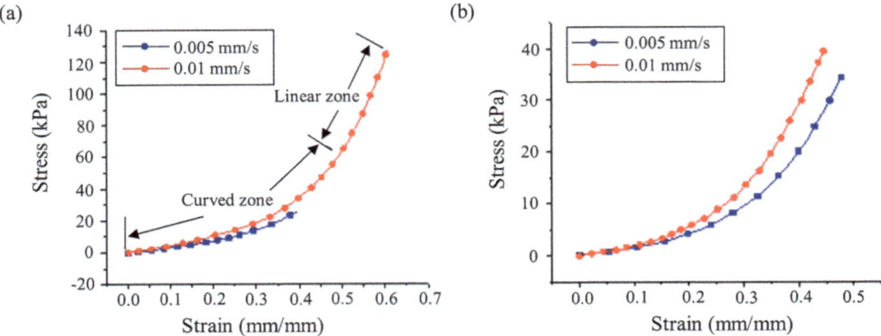

**Fig. 4.26** Effect of displacement rates on stress vs. strain curves of (**a**) 4% and (**b**) 5% alginate specimens under static compression loading

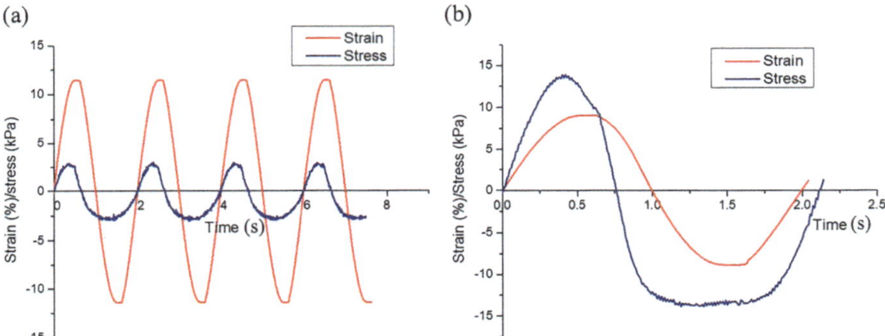

**Fig. 4.27** Time-dependent stress and strain waves of (**a**) 4% alginate and (**b**) 5% alginate specimens

viscous response and the linear zone represents the elastic response. The viscous response of the material can be quantified in terms of the radius of curvature of the curved region, whereas the elastic response can be analyzed in terms of the elastic modulus of the linear region. Figure 4.26 shows the elastic response of the material in terms of the elastic modulus increases with the displacement rate, but the radius of curvature decreases. This indicates the alginate gels behave more elastically at fast displacement rates and more viscously at slow displacement rates.

The stress response of the 4% and 5% alginate materials was recorded, which is shown in Fig. 4.27 along with the applied sinusoidal strain. The sinusoidal strains applied on both 4% and 5% alginate specimens have a frequency of 3.14 radians/s (or a time period ($T$) of 2 s as $\omega = 2\pi/T$); their amplitudes are 0.114 and 0.0897 mm/mm, respectively. Thus, the sinusoidal strains applied on the 4% and 5% alginate specimens are given by, respectively:

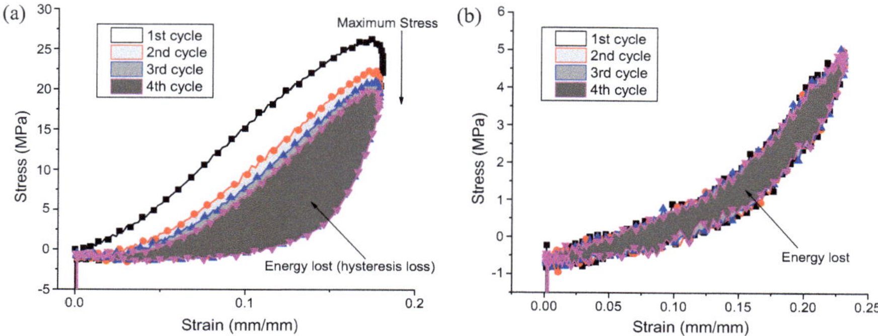

**Fig. 4.28** Stress-strain curve for cyclic loading for the (**a**) 5% alginate and (**b**) 4% alginate specimens

$$\varepsilon = 0.114 \sin(3.14t), \tag{4.51}$$

$$\varepsilon = 0.0897 \sin(3.14t). \tag{4.52}$$

By fitting the data shown in Fig. 4.27, the stress wave equations were obtained for the 4% and 5% alginate specimens, respectively:

$$\sigma = 2.91 \sin(3.14t + 1.07), \tag{4.53}$$

$$\sigma = 13.89 \sin(3.14t + 0.72). \tag{4.54}$$

The hysteresis behavior of the alginate specimens under cyclic loading is shown in Fig. 4.28.

The stress-strain curves in Fig. 4.28 show the loop of the curve decreases after each successive cycle. The area of each loop represents the energy lost during that cycle. The figure shows the energy loss decreases after each cycle for both specimens. This energy loss is called hysteresis loss in the case of viscoelastic materials. The amount of energy lost for both alginate materials was found to continuously decrease until it reached zero. This means the viscous response of the material continually decreases after each cycle until only the elastic response remains. One more interesting feature observed in the case of 5% alginate gel is that the value of maximum stress continuously decreases (Fig. 4.28a). This behavior was also evident in the 4% alginate specimen after certain cycles. This decreasing stress and energy loss after successive cycles is known as preconditioning behavior of a viscoelastic material. This behavior has been noted for many biological materials by different researchers.

## 4.5   Mechanical Properties of Scaffolds

Tissue scaffolds are designed in the form of three-dimensional porous structures to act as templates that provide the physical environment for tissue formation. To effectively facilitate tissue regeneration, scaffolds should provide the required mechanical support and appropriate environment for tissue regeneration. The mechanical properties of tissue scaffolds are affected by structural parameters (such as pore geometry, strand size, strand spacing, and strand orientation) and the material from which they are made. The mechanical properties of tissue scaffolds are also time dependent due to their continuous degradation after implantation. These factors associated with the mechanical properties of tissue scaffolds are discussed in the following sections.

### 4.5.1   Influence of Scaffold Structure

Variation in pore geometry configuration in terms of the strand diameter (D), strand size (SZ), strand spacing (SS), and strand orientation (SO) changes the percent porosity and thus affects the mechanical properties of scaffolds. The effect of scaffold structure on mechanical properties can be demonstrated with the help of a previous study [4] in which scaffolds were made from PCL (with a molecular weight 80,000 $g \cdot mol^{-1}$) with varying pore geometries as listed in Table 4.3. The percent porosity for each scaffold was measured and demonstrated dependence on the pore geometry.

The scaffolds were tested under both compressive and tensile loading, with the measured compressive ($E_C$) and tensile ($E_T$) moduli of the scaffolds along with their porosities listed in Table 4.4. The effect of porosity on the elastic moduli of the tissue scaffolds was analyzed using linear regression analysis of scaffold porosity vs. tensile strength and vs. compressive elastic modulus. The linear regression plots for these two cases are shown in Fig. 4.29a, b, respectively. The resulting linear regression equations, Pearson's coefficient (r), $R^2$ values, and $p$-values are reported in Table 4.5.

The linear regression analysis results show scaffold porosity is negatively correlated with both the compressive and tensile moduli of the scaffolds. The linear regression plots show in Fig. 4.30a, b indicate the compressive and tensile elastic modulus of the scaffolds both decrease with increasing porosity. The outcomes of this case study show that variations in structural parameters affect the mechanical properties of scaffolds by changing the percent porosity.

**Table 4.3** Scaffolds made with varying geometries and their measured porosities

| Group no. | SZ (mm) | SO (degrees) | SS (mm) | % porosity |
|---|---|---|---|---|
| 1 | 400 | 0-90 | 1 | 36.91 |
| 2 | 300 | 0-45 | 1 | 42.65 |
| 3 | 300 | 0-45-90-135 | 1 | 44.53 |
| 4 | 200 | 0-90 | 1 | 56.84 |
| 5 | 400 | 0-45 | 1.5 | 51.98 |
| 6 | 400 | 0-45-90-135 | 1.5 | 53.53 |
| 7 | 400 | 0-90 | 2 | 63.72 |
| 8 | 300 | 0-90 | 1.5 | 67.39 |
| 9 | 300 | 0-45-90-135 | 2 | 68.73 |
| 10 | 300 | 0-45 | 2 | 70.65 |
| 11 | 200 | 0-45 | 1.5 | 72.13 |
| 12 | 200 | 0-45-90-135 | 1.5 | 75.98 |
| 13 | 200 | 0-90 | 2 | 80.28 |

**Table 4.4** Compressive and tensile moduli of different groups of scaffolds with different percent porosity values

| Group no. | % porosity | Compressive modulus ($E_C$) in MPa | Tensile modulus ($E_T$) in MPa |
|---|---|---|---|
| 1 | 36.91 | 56.46 | 46.04 |
| 2 | 42.65 | 39.23 | 20.83 |
| 3 | 44.53 | 33.62 | 25.36 |
| 4 | 56.84 | 28.79 | 18.15 |
| 5 | 51.98 | 17.16 | 24.75 |
| 6 | 53.53 | 15.59 | 34.4 |
| 7 | 63.72 | 13.31 | 15.86 |
| 8 | 67.39 | 14.81 | 9.74 |
| 9 | 68.73 | 11.43 | 16.41 |
| 10 | 70.65 | 11.47 | 11.68 |
| 11 | 72.13 | 8.68 | 9.66 |
| 12 | 75.98 | 8.69 | 14.59 |
| 13 | 80.28 | 6.63 | 6.03 |

## *4.5.2  Influence of Scaffold Materials*

The mechanical properties of tissue engineering scaffolds depend not only on structural properties as discussed in the previous section but also on the material used for fabrication. Material selection for a desired scaffold design depends on the mechanical properties of the native tissue into which the scaffold is to be transplanted to facilitate tissue growth. This section presents the effects of material selection on the mechanical properties of porous scaffolds, based on the results of a study of the effects of scaffold material on compressive properties [4].

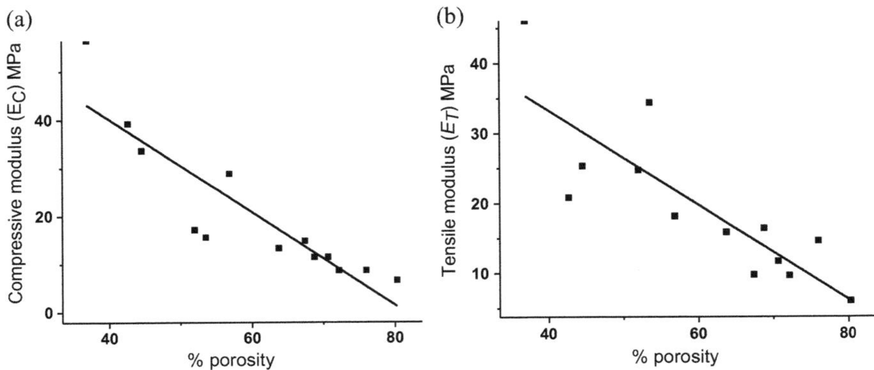

**Fig. 4.29** Linear regression plots showing the effect of percent porosity on (**a**) compressive and (**b**) tensile moduli of PCL scaffolds

**Table 4.5** Results of the linear regression analysis between scaffold porosity and compressive as well as tensile moduli

| Independent parameter | Dependent parameter | Linear regression equation | Pearson's coefficient ($r$) | $R^2$ value | $p$-value |
|---|---|---|---|---|---|
| % porosity | Compressive modulus ($E_C$) | $E_C = 78.82 - 0.97\ (\%porosity)$ | −0.89 | 0.80 | 0.001 |
| % porosity | Tensile modulus ($E_T$) | $E_T = 60.35 - 0.68\ (\%porosity)$ | −0.84 | 0.70 | 0.001 |

PCL materials with different molecular weights (80,000, 45,000, and 10,000 g/mol, identified as PCL 80, PCL 45, and PCL 10, respectively) were selected to fabricate porous scaffolds with similar structural parameters. The CAD model of a bulk/block PCL scaffold was designed using Magics Envisiontec, as discussed in Case 2 of Sect. 4.4.3. The 3D model of the scaffold had a cubic geometry with dimensions of $15 \times 15 \times 2.4$ mm$^3$, 10 layers, a 0.26-mm strand thickness, 1-mm interstrand spacing, and a 0–90° pattern.

The scaffolds designed from the PCL 10, PCL 45, and PCL 80 materials were subjected to compressive loading on an Instron 3366 material testing machine. The stress-strain curves obtained from compressive testing were recorded and the values obtained for yield strength, compressive strength, and compressive modulus were calculated and compared. The stress-strain curves obtained for these scaffolds are shown in Fig. 4.30.

The compressive modulus, yield strength, and compressive strength of these scaffolds were calculated from the stress-strain curves obtained from the compression tests. The values of these mechanical properties for the different scaffolds were compared to analyze the effect of PCL molecular weight on compressive behavior. The statistical analysis shows the various compressive properties of the scaffolds fabricated with PCL 10, PCL 45, and PCL 80 significantly differed. Graphical illustrations of yield strength and compressive strength depicting the significant

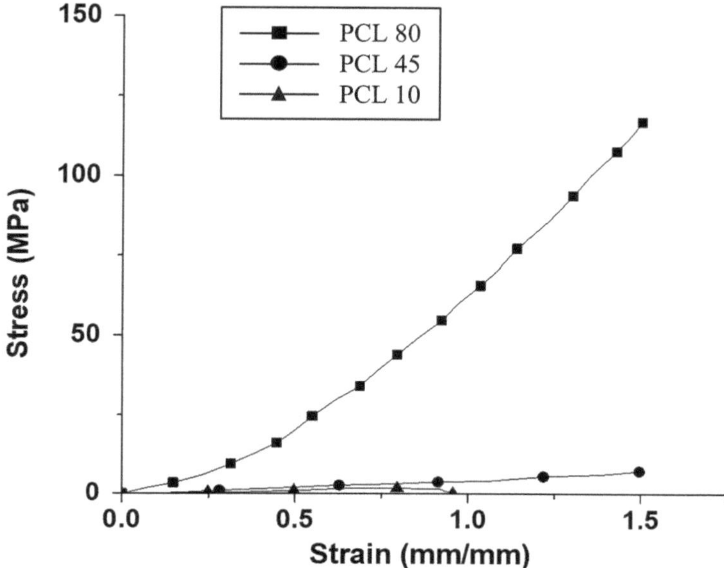


**Fig. 4.30** Stress-strain curves obtained for scaffolds made of PCL materials with different molecular weights [4]

difference in these properties for the PCL 10, PCL 45, and PCL 80 scaffolds are shown in Fig. 4.31a, b, respectively. The scaffold fabricated with PCL 10 had the lowest values of all compressive properties. This scaffold was very brittle with the lowest toughness and reached the failure value very quickly compared to the PCL 45 and PCL 80 scaffolds. The compressive properties of the PCL 80 scaffold were the highest. These results show that scaffolds made from PCL 10 cannot be used to facilitate tissue growth in load-bearing situations. This case study shows the material used for scaffold fabrication significantly affects the resulting mechanical properties, and therefore, the material for scaffold design should be carefully selected for particular tissue engineering applications.

## 4.5.3   Time-Dependent Mechanical Properties

The mechanical properties of a biomaterial scaffold are not constant but change with time. Once a scaffold is implanted to facilitate tissue regeneration at the desired site, material degradation begins along with the regeneration of new tissue. Due to the material degradation, the mechanical strength and stiffness of the scaffold degrade with time. For example, the effect of material degradation on the scaffold elastic modulus has been studied [6]. The scaffolds for this study were fabricated from PCL mixed with hydroxyapatite with a strand spacing of 700 µm and elliptical

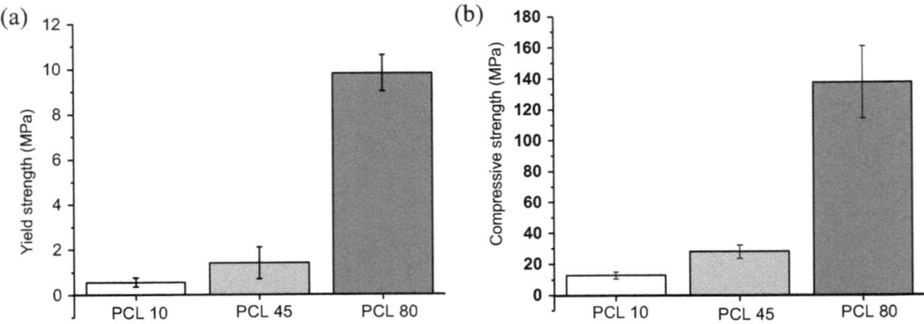

**Fig. 4.31** Diagrams showing the values of (**a**) yield strength and (**b**) compressive strength of scaffolds fabricated from PCL 10, PCL 45, and PCL 80 [4]

cross-section major and minor radii of 150 and 147 µm, respectively. To examine the effect of degradation, these scaffolds were incubated in PBS at 40 °C and subjected to compression testing at 36% relative humidity on an Instron 1020 machine at a displacement rate of 2 mm/min at four different time points (t = 1, 6, 13, and 18 weeks). The elastic modulus of these scaffolds was found to drastically deteriorate with time from about 31 MPa at 1 week to 12 MPa at 18 weeks. As such, success in scaffold implantation requires that the rate of scaffold degradation and tissue regeneration match such that the declining strength of the scaffold is compensated by the increasing strength provided by the regenerating tissue.

## 4.6  Methods to Improve the Mechanical Properties of Scaffolds

### 4.6.1  Use of Composite Materials

Composite scaffolds are developed by exploiting the complementary mechanical, chemical, and biological properties of various materials. These scaffolds are designed to have improved functionality in terms of mechanical strength, bioactivity, surface reactivity, and drug or growth factor delivery capability. Biological materials such as collagen, proteoglycans, alginate-based substrates, and chitosan are considered to be biologically active as these materials promote excellent cell adhesion and growth. On the other hand, synthetic polymers such as polystyrene, poly-L-lactic acid (PLLA), polyglycolic acid (PGA), and poly-DL-lactic-co-glycolic acid can be easily fabricated with a tailored architecture. The degradation characteristics of synthetic polymers can also be controlled by changing their composition. Moreover, both natural and synthetic polymers have some drawbacks that can be overcome by combining them to produce natural/synthetic polymer composite scaffolds. For example, the bioactive properties of PCL materials can be improved through combination with other polymers such as gelatin and chitosan [7]. The degradable fragments of PLLA polymers have high crystallinity and lead to inflammation in the body, but this problem is reduced in PLLA/ginsenoside (Rg3) composite scaffolds [8]. Silk composites are also very commonly used to achieve improved properties in terms of elasticity, degradation rate, and porosity. Composite scaffolds for load-bearing applications are designed by combining the strength, stiffness, and osteoconductivity of ceramics with the flexibility, toughness, resorbability, and processability of polymers to achieve an optimum outcome in terms of mechanical compatibility. Commercially available natural polymer-ceramic composites consisting of type I collagen and calcium phosphate minerals can mimic native bone tissue [9].

Natural polymer-based composites can be divided into four categories: (1) collagen-based composites, (2) gelatin-based composites, (3) chitosan-based composites, and (4) bacterially derived polymer composites. A brief summary of these composites from previous studies including fabrication methods, porosities, and mechanical properties is reported in Table 4.6 [9].

The use of composite scaffolds to achieve improved mechanical properties can be demonstrated with a case study based on natural polymer-based scaffolds [10]. The 3D scaffolds were prepared using microfibrillated bacterial cellulose (MFC) and poly(3 hydroxybutyrate) (P(3HB)) materials employing a novel compression molding/particulate leaching technique. These scaffolds were prepared with a size of $0.5 \times 0.5 \times 0.5$ cm$^3$ and various MFC contents (10, 20, 30, 40, and 50 wt%) and subjected to compression testing. The stress-strain curves were recorded and the compressive modulus and yield strength determined. The compressive modulus of the composite scaffolds was found to be higher than a neat P(3HB) scaffold by 35, 37, 64, and 124% for MFC contents of 10, 20, 30, and 40 wt%, respectively. The compressive yield strength of the composite scaffolds was also found to progressively increase with the addition of MFC to the polymer matrix. The study shows the mechanical properties of tissue engineering scaffolds can be significantly improved by developing composite scaffolds comprised of more than one substrate.

The mechanical properties of composite scaffolds made from PCL and poly(lactic acid) (PLA) as well as PCL and hyaluronic acid (HA) have also been examined [11]. Hybrid spools of PCL-PLA and PCL-HA were made using a Randcastle microfilament extruder. These spools were further loaded into a commercially available 3D printer to make log-pile scaffolds, which were produced in two different sizes: $10.5 \times 10.5 \times 12$ mm$^3$ and 6 mm$^3$. Compression tests were performed on a materials test systems (MTS) machine to compare their mechanical properties. Compressive modulus values obtained from the stress-strain curves were 159.2, 59.3, and 44.3 MPa for the PCL-PLA, PCL-HA, and PCL scaffolds, respectively. The compressive strength of the PCL-PLA scaffold was also found to be higher compared to the PCL-HA and PCL scaffolds.

**Table 4.6** Summary of different composite scaffolds in terms of fabrication method and mechanical properties

| Composite | Method | Elastic modulus | Tensile strength | Compressive modulus | Compressive strength | % porosity |
|---|---|---|---|---|---|---|
| Col-nHA | Freeze drying | – | – | 5.5 kPa | – | 99.4 |
| Col-nHA | Immersion method | – | – | 4 kPa | – | 98.9 |
| Col-PLGA-nHA | Layer-by-layer solvent casting | 1.2 GPa | 9.7 MPa | – | – | – |
| Gel-nHA | Electrospinning | 412 MPa | 4.4 MPa | – | – | – |
| Gel-PCL-nHA | Layer-by-layer solvent casting | 23.5 MPa | 3.7 MPa | – | – | – |
| Gel-mHA | Freeze drying | – | – | 4.5 MPa | 0.4 MPa | – |
| CS-nHA | In situ precipitation | 704 MPa | – | – | 230 MPa | – |
| CS-nBG | Solvent casting | 2.6 GPa | 49.6 MPa | – | – | – |
| CS-PLA-nHA | In situ precipitation | 880 MPa | – | – | 266 MPa | 85 |
| CS-mHA | Rapid prototyping | 16 GPa | – | – | – | 50 |
| CS-mHA | Solvent casting | 3.3 GPa | 42 MPa | – | – | – |
| PHBV-nHA | Thermally induced phase separation | – | – | 10 MPa | 2.9 MPa | 94.9 |
| SF-mHA | Solvent casting | – | – | 1 MPa | – | – |

*Col* Type I collagen, *nHA* nano hyaluronic acid, *PLGA* poly(lactic-co-glycolic acid), *Gel* gelatin, *mHA* microporous hyaluronic acid, *CS* chitosan, *nBG* nano bioactive glass, *PLA* poly(lactic acid), *PHBV* polyhydroxybutyrate-co-(3-hydroxyvalerate), *SF* silk fibroin

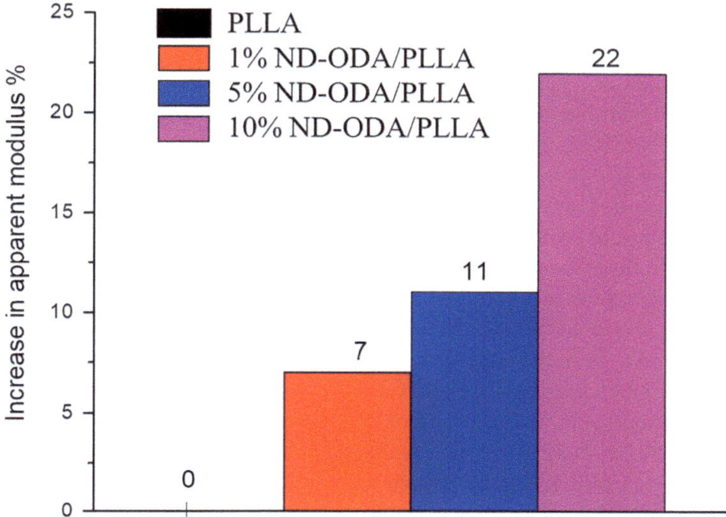


**Fig. 4.32** Improvement of apparent modulus in compression with the addition of octadecylamine-functionalized nanodiamond (ND-ODA) in a poly(L-lactic acid) (PLLA) scaffold [12]

## 4.6.2   Addition of Fillers

The mechanical properties of scaffolds can be effectively improved by incorporating fillers such as nanoparticles into scaffold materials. Nanodiamond is a nanoparticle with very good physical and chemical properties that can be used to increase the mechanical properties of scaffolds for bone tissue engineering applications. The important properties of nanodiamonds are their nanoscale size, nearly spherical shape, rich surface chemistry, excellent biocompatibility, physico-chemical characteristics, and low cytotoxicity. The adequate dispersion of nanodiamonds can improve the strength, toughness, and thermal stability of nanocomposites. The effect of the addition of nanodiamond to a PLLA scaffold is shown in Fig. 4.32 [12], with the compressive modulus of the PLLA scaffolds increasing proportionally to the addition of 1, 5, and 10 wt.% octadecylamine-functionalized nanodiamond (ND-ODA). The failure strain of these scaffolds increased 280% and the fracture energy in tension increased 310% by adding 10 wt.% ND-ODA. Other nanoparticles that are effectively used to improve the mechanical properties of scaffolds are hydroxyapatite, nanoSiO$_2$ and MgO particles, bioactive glass particles, and silver nanoparticles [12].

### *4.6.3  Hybrid Structures*

Hybrid scaffolds are fabricated from two or more different, but complementary, materials. For example, a solid polyester material and hydrogel can be combined to form a hybrid scaffold, where the solid material provides mechanical strength while the hydrogel provides a cell-supportive matrix, as discussed in Chap. 2. Notably, hybrid scaffolds can also be created using different approaches.

The infiltration of hydrogel into a solid scaffold is a very popular method in tissue engineering. In this process, a solid scaffold framework is loaded with hydrogel; this process is as simple as dripping the hydrogel onto the solid scaffold or seeding the solid scaffold with a hydrogel cell suspension [13]. Researchers have also successfully hybridized a synthetic poly (L-lactide-co-ε-caprolactone) (PLCL) polymer with chondrocyte-embedded fibrin gels (FG) and HA hydrogel [14]. They developed a porous solid framework with the help of press molding, salt leaching, and then freeze drying of the PLCL-NaCl mixture. The chondrocyte hydrogel was quickly incubated in the PLCL scaffold just after preparation. These scaffolds were implanted and observed in vivo. A well-developed, homogeneously distributed cartilage construct was obtained after 8 weeks, and had compressive properties comparable to natural articular cartilage. In another study [15], a hybrid construct was developed using a synthetic-polymer printed framework into which cells were then impregnated. These types of 3D printed and cell-impregnated hybrid constructs are promising in terms of mimicking the structural and biological features of cartilage.

The other approach used to provide uniform infiltration of hydrogels into solid scaffolds is vacuum-assisted infusion. The vacuum-based infusion technique has been used to hybridize an agarose/fibrin hydrogel with 3D woven PCL or PGA scaffolds [16]. The initial mechanical properties of this hybrid scaffold were improved in this way by designing a woven reinforcing component. This hybrid scaffold also had biomimetic mechanical properties in terms of anisotropy, viscoelasticity, and tension-compression nonlinearity. The viscoelastic creep behavior and stiffness were also improved by incorporating the hydrogel component. The hybrid woven PGA/PCL-agarose fibrin scaffold designed in this study is shown in Fig. 4.33.

A hybrid scaffold structure can also be created using two or more fabrication techniques. For example, one study used printing as well as electrospinning fabrication techniques to produce hybrid scaffolds of PCL and collagen biomaterials [17]. Using the printing technique, two layers of perpendicular strands were printed from the melted PCL and then collagen nanofibers were electrospun on top of the PCL strands. A 3D hybrid scaffold was fabricated by repeating this pattern and later these scaffolds were seeded with cells. The mechanical properties of the hybrid scaffolds were assessed using a tensile test, with the elastic modulus and biological activity found to be better than a pure PCL scaffold.

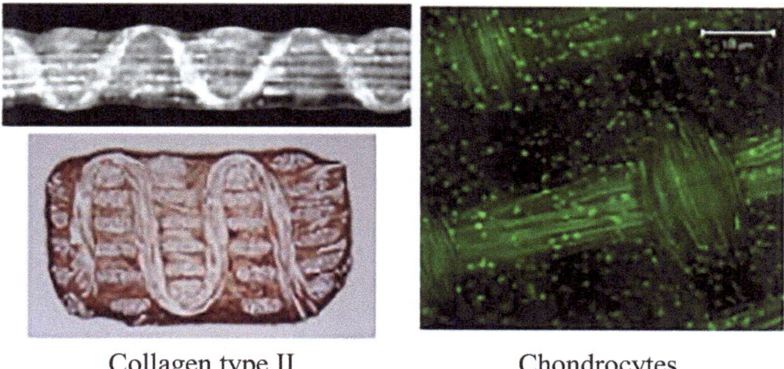

Collagen type II          Chondrocytes

**Fig. 4.33** A hybrid woven PGA/PCL-agarose fibrin scaffold [16]

## 4.7 Summary

Tissue scaffolds provide 3D structural and mechanical support for cell growth and tissue regeneration to aid in the repair of damaged tissues. For this, the mechanical properties of scaffolds play a crucial role and ideally should match those of healthy tissues at the native site. This requires the mechanical properties of both the native tissue and scaffold to be measured and characterized.

The mechanical properties of native tissues and scaffolds can be characterized through mechanical testing. During mechanical testing, a specimen or sample is subjected to loading conditions of tensile/compressive force, bending moment, and/or torque while the deformation of the specimen is measured or recorded. Stress and strain are two important concepts associated with the applied force and the deformation. The stress-strain curve varies with the nature of the sample being tested and can be used to determine the mechanical property parameters, including the elastic modulus, yield stress/strain, ultimate stress/strain, and failure stress/strain. These mechanical properties are static and measured by means of static testing or applying loading or displacement with a slow rate of change.

Viscoelasticity is a dynamic behavior of materials that exhibit both viscous and elastic characteristics when undergoing deformation. Viscoelastic properties are measured and characterized by means of dynamic testing methods where the rates of loading or displacement affect the testing results. The dynamic testing methods include cyclic, creep, stress-relaxation, or frequency-dependent testing, which are respectively characterized by the area inside the hysteresis loop, parameter of creep $(\tau_{3/2})$, half-stress relaxation time $(\tau_{1/2})$, and complex modulus $E(\omega)$. The linear viscoelastic behavior can be modeled using a combination of spring and dashpot elements. Among such models, the Maxwell and Kelvin-Voigt models are the most common. The Maxwell model consists of a spring and a dashpot in series and is good for representing stress relaxation behavior, while the Kelvin-Voigt model consists of a spring and a dashpot in parallel and is good for representing creep behavior.

The text provides some basic considerations regarding mechanical property measurements of native tissues and scaffolds. Specimen preparation is an important step, and during mechanical testing, a number of testing conditions can affect the measured results, such as sample mounting, operating temperature and humidity, and loading directions. Key steps to measure the mechanical properties of native tissue and printed scaffolds, including sample preparation, testing procedure, and evaluation of the important mechanical properties from the results obtained, were discussed in three case studies.

The mechanical properties of a printed scaffold can be influenced by its structure and the material(s) from which it is fabricated. The mechanical properties of tissue scaffolds are also not constant but change with time due to the dynamics of both material degradation and tissue regeneration. Different methods can be employed to improve the mechanical properties of scaffolds including the use of composite materials, additional fillers, and hybrid structures.

**Problems**

1. Briefly explain why the mechanical properties of scaffolds are important in scaffold-based tissue engineering.
2. A biomaterial sample is 30 mm long and has a square cross-sectional area of $5 \text{ mm} \times 5 \text{ mm} = 25 \text{ mm}^2$. A downward force of 20 N results in a deflection of 0.1 mm at the beam center. Calculate the elastic modulus assuming the material behavior is linear.
3. The following tensile load displacement data are for a ductile material with a gauge length of 13 mm and an initial cross-sectional area of $25 \text{ mm}^2$. Find the Young's modulus, yield strength, and ultimate strength of the material.

| Length (mm) | Load (N) |
| --- | --- |
| 13 | 0 |
| 13.01 | 1.01 |
| 13.02 | 5.13 |
| 13.03 | 9.1 |
| 13.04 | 13.22 |
| 13.05 | 17.18 |
| 13.06 | 21.1 |
| 13.07 | 22.42 |
| 13.08 | 23.03 |
| 13.09 | 23.56 |
| 13.1 | 24.06 |
| 13.11 | 24.56 |
| 13.12 | 25.06 |
| 13.13 | 25.57 |
| 13.14 | 25.97 |
| 13.15 | 26.4 |
| 13.16 | 26.81 |
| 13.17 | 27.21 |

(continued)

| Length (mm) | Load (N) |
| --- | --- |
| 13.18 | 27.58 |
| 13.19 | 27.87 |
| 13.2 | 28.17 |
| 13.21 | 28.48 |
| 13.22 | 28.82 |
| 13.23 | 29.12 |
| 13.24 | 29.42 |
| 13.25 | 29.71 |
| 13.26 | 29.96 |
| 13.27 | 30.16 |
| 13.28 | 0.06 |
| 13.29 | 0.04 |
| 13.3 | 0.05 |
| 13.31 | 0.05 |

4. A rectangular beam type specimen with dimensions of $2 \times 2 \times 50$ mm$^3$ was prepared from a horse femur bone for a three-point bend test. The values of flexural strength and flexural elastic modulus obtained from the bending test were 204 MPa and 17.1 GPa, respectively. Calculate (a) the maximum load sustained by the specimen before fracture and (b) the slope of the initial load-displacement curve. If the flexural strain value of the bone specimen was 0.0032 mm/mm, also calculate the maximum deflection of the mid-point of the specimen under bending. The span length of the specimen can be taken as 40 mm.

5. Two 1-mm diameter cylindrical specimens of cancellous bone tissue were subjected to 5 and 10 N-mm of torque, respectively, during a torsional test. Determine the maximum shear stresses developed in the bone specimens. If the shear stress developed in one of the specimens corresponding to $5.5 \times 10^{-4}$ radians/mm angle of twist per unit length is 10 MPa, calculate its torsional modulus.

6. A sample has an initial length of 10 mm and a cross-sectional area of 25 mm$^2$. An initial strain is applied to the sample, then followed by a sinusoid strain loading. The following stress-strain data are recorded during testing.

| Time (s) | Stress (MPa) | Strain (mm/mm) |
| --- | --- | --- |
| 0 | 2.644 | 1.000 |
| 0.01 | 2.971 | 1.294 |
| 0.02 | 2.926 | 1.476 |
| 0.03 | 2.528 | 1.476 |
| 0.04 | 1.928 | 1.294 |
| 0.05 | 1.356 | 1.000 |
| 0.06 | 1.029 | 0.706 |
| 0.07 | 1.074 | 0.524 |
| 0.08 | 1.472 | 0.524 |

(continued)

| Time (s) | Stress (MPa) | Strain (mm/mm) |
| --- | --- | --- |
| 0.09 | 2.072 | 0.706 |
| 0.1 | 2.644 | 1.000 |
| 0.11 | 2.971 | 1.294 |
| 0.12 | 2.926 | 1.476 |
| 0.13 | 2.528 | 1.476 |
| 0.14 | 1.928 | 1.294 |
| 0.15 | 1.356 | 1.000 |

(1) Fit a sine wave model to the strain data to find the angular frequency of the strain loading in rads/s and convert it to a frequency. (2) Determine the value of loss angle, $\delta$, by plotting stress and strain as functions of radians. (3) Find the loss modulus and the storage modulus.

7. Name one mechanical testing method that can be used to measure the static properties of materials, perform a literature review on the use of this testing method to measure the mechanical properties of a tissue or scaffold by briefly presenting the sample preparation and mechanic-property characterization.

8. Name one testing method that can be used to measure the viscoelastic properties of materials, and perform a literature review on the use of this testing method to measure the viscoelastic properties of a tissue or scaffold by briefly presenting the sample preparation and viscoelastic-property characterization.

9. Briefly explain how the mechanical properties of a scaffold can be designed or adjusted as desired.

# References


1. Y. Ma, T. Han, Q. Yang, et al., Viscoelastic cell microenvironment: Hydrogel-based strategy for recapitulating dynamic ECM mechanics. Adv. Funct. Mater. **31**(24), 2100848 (2021). https://doi.org/10.1002/adfm.202100848

2. Z. Yazdanpanah, J.D. Johnston, D.M.L. Cooper, X. Chen, 3D bioprinted scaffolds for bone tissue engineering: State-of-the-art and emerging technologies. Front. Bioeng. Biotechnol. **10** (2022). https://doi.org/10.3389/fbioe.2022.824156

3. N.K. Sharma, J. Nayak, D.K. Sehgal, R.K. Pandey, Studies on post-yield behavior of cortical bone. Appl. Mech. Mater. **232**, 157–161 (2012). https://doi.org/10.4028/www.scientific.net/AMM.232.157

4. A.D. Olubamiji, Z. Izadifar, J.L. Si, et al., Modulating mechanical behaviour of 3D-printed cartilage-mimetic PCL scaffolds: Influence of molecular weight and pore geometry. Biofabrication **8**(2), 025020 (2016). https://doi.org/10.1088/1758-5090/8/2/025020

5. N.K. Sharma, On the viscoelastic properties of alginate/nHA composite hydrogels for potential bone tissue engineering. Dissertation, University of Saskatchewan (2023)

6. N.K. Bawolin, M.G. Li, X.B. Chen, W.J. Zhang, Modeling material-degradation-induced elastic property of tissue engineering scaffolds. J. Biomech. Eng. **132**(11), 111001 (2010). https://doi.org/10.1115/1.4002551

7. Y. Zhang, H. Ouyang, C.T. Lim, et al., Electrospinning of gelatin fibers and gelatin/PCL composite fibrous scaffolds. J. Biomed. Mater. Res. Part B Appl. Biomater. **72B**, 156–165 (2005). https://doi.org/10.1002/jbm.b.30128

8. S. Stratton, N.B. Shelke, K. Hoshino, et al., Bioactive polymeric scaffolds for tissue engineering. Bioact. Mater. **1**, 93–108 (2016). https://doi.org/10.1016/j.bioactmat.2016.11.001

9. R. Yunus Basha, T.S. SK, M. Doble, Design of biocomposite materials for bone tissue regeneration. Mater. Sci. Eng. C **57**, 452–463 (2015). https://doi.org/10.1016/j.msec.2015.07.016

10. E. Akaraonye, J. Filip, M. Safarikova, et al., Composite scaffolds for cartilage tissue engineering based on natural polymers of bacterial origin, thermoplastic poly(3-hydroxybutyrate) and micro-fibrillated bacterial cellulose. Polym. Int. **65**, 780–791 (2016). https://doi.org/10.1002/pi.5103

11. L.D. Albrecht, S.W. Sawyer, P. Soman, Developing 3D scaffolds in the field of tissue engineering to treat complex bone defects. 3D Print Addit. Manuf. **3**, 106–112 (2016). https://doi.org/10.1089/3dp.2016.0006

12. J. Corona-Gomez, X. Chen, Q. Yang, Effect of nanoparticle incorporation and surface coating on mechanical properties of bone scaffolds: A brief review. J. Funct. Biomater. **7**, 18 (2016). https://doi.org/10.3390/jfb7030018

13. Z. Izadifar, X. Chen, W. Kulyk, Strategic design and fabrication of engineered scaffolds for articular cartilage repair. J. Funct. Biomater. **3**, 799–838 (2012). https://doi.org/10.3390/jfb3040799

14. Y. Jung, S.-H. Kim, Y.H. Kim, S.H. Kim, The effect of hybridization of hydrogels and poly (L-lactide-co-ε-caprolactone) scaffolds on cartilage tissue engineering. J. Biomater. Sci. Polym. Ed. **21**, 581–592 (2010). https://doi.org/10.1163/156856209X430579

15. Z. Izadifar, T. Chang, W. Kulyk, et al., Analyzing biological performance of 3D-printed, cell-impregnated hybrid constructs for cartilage tissue engineering. Tissue Eng. Part C Methods **22**, 173–188 (2016). https://doi.org/10.1089/ten.tec.2015.0307

16. F.T. Moutos, Biomimetic composite scaffolds for the functional tissue engineering of articular cartilage. Dissertation, Duke University (2009)

17. H. Lee, M. Yeo, S. Ahn, et al., Designed hybrid scaffolds consisting of polycaprolactone microstrands and electrospun collagen-nanofibers for bone tissue regeneration. J. Biomed. Mater. Res. Part B Appl. Biomater. **97B**, 263–270 (2011). https://doi.org/10.1002/jbm.b.31809

# Chapter 5
# Preparation of Biomaterial Solutions and Characterization of Their Flow Behavior

## 5.1  Introduction

To print scaffolds, the selected biomaterials and associated biological elements, such as living cells and biomolecules, should be prepared in the form of a liquid or solution. More specifically, the biomaterial solutions or bio-ink must be prepared to have appropriate flow behaviors such that they can be extruded to form scaffolds with a 3D structure. As such, the preparation of biomaterial solutions is a critical step in printing tissue scaffolds with desired structural and functional properties. This chapter discusses the basics of biomaterial solution preparation (including concentration and sterilization) and then describes the preparation of solutions from various biomaterials for bioprinting and the methods/techniques used to characterize their flow behavior.

## 5.2  Preparation of Biomaterial Solutions

### 5.2.1  Basics of Solution Preparation

The biomaterials selected for scaffold bioprinting must be dissolved to form solutions or solution-like phases with appropriate flow behavior. Biomaterial solutions with highly viscous flow behavior require a larger extruding force during the printing process but more easily form robust printed scaffold structures; on the other hand, low-viscosity solutions require less extruding force but may form structures that are unstable or prone to collapse. As such, the preparation of biomaterial solutions with flow behaviors appropriate for the bioprinting process is of great importance. Notably, most available biomaterials used in bioprinting do not come in a fluid form but instead as gels, powders, or particles that need to be dissolved in solvents under certain conditions before use. Generally, these materials, as solutes, can be



D. X. B. Chen, *Extrusion Bioprinting of Scaffolds for Tissue Engineering*,
https://doi.org/10.1007/978-3-031-72471-8_5

classified as either water soluble or non-soluble. A water-soluble material can be directly dissolved into water or water-based solutions. Widely used biomaterials in scaffold bioprinting, such as alginate and gelatin, are water-soluble materials and can dissolve in water in times ranging from minutes to hours [1]. Many materials also barely dissolve in water under conditions of neutral pH and room temperature but can easily dissolve at non-neutral pH or elevated temperature. These materials, such as chitosan, collagen, and polyvinyl alcohol (PVA), are also classified as water-soluble materials. Non-water-soluble biomaterials that are utilized in extrusion-based bioprinting for scaffold fabrication must be processed into solution-like phases before application. To achieve this, these non-soluble materials are either dissolved in special organic solvents (e.g., chloroform for polycaprolactone (PCL)) or thermally melted by controlling the temperature during material extrusion [2]. Table 5.1 summarizes common biomaterials used in scaffold bioprinting.

The concentration of a biomaterial solution represents the amount of biomaterial (as a solute) dissolved in a certain amount of solution. Concentration values are often

**Table 5.1** Common water soluble and non-soluble scaffold materials

| Scaffold materials | Water soluble or non-soluble | Typical preparation conditions |
| --- | --- | --- |
| Alginate | Water soluble | Dissolves within hours in water or water-based solutions at room temperature |
| Chitosan | Water soluble | Dissolves within hours in weak acid at room temperature |
| Agarose | Water soluble | Easily dissolves in near-boiling water or water-based solutions |
| Hyaluronic acid (HA) | Water soluble | Dissolves within hours in weak acid at room temperature |
| Collagen | Water soluble | Dissolves in weak acid at room temperature and gels at neutral pH at 37 °C |
| Gelatin | Water soluble | Dissolves in water or water-based solutions at 37 °C and gradually gels as temperature drops |
| Fibrin | Water soluble | Dissolves within minutes in water or water-based solutions at room temperature |
| Poly(ethylene glycol) (PEG) | Water soluble | Dissolves in water or water-based solutions at room temperature and polar solvents such as acetone |
| Poly(ethylene oxide) (PEO) | Water soluble | Dissolves in water or water-based solutions at room temperature and polar solvents such as acetone |
| Polyvinyl alcohol (PVA) | Water soluble | Dissolves within minutes in water or water-based solutions over 40 °C |
| Decellularized matrix (dECM) | Water soluble | Dissolves in water or water-based solutions |
| Polycaprolactone (PCL) | Non-soluble | Dissolves in organic solvent such as chloroform; melts at 60 °C |
| Polylactic acid (PLA) | Non-soluble | Dissolves in organic solvents such as propanol; melts at 180 °C |
| Polyglycolic acid (PGA) | Non-soluble | Dissolves in solvents such as hexafluoroisopropanol; melts at 225 °C |

expressed in terms of relative units, most commonly mass percent, volume percent, and mass/volume percent. The mass percent (%w/w) is the percentage of solute mass in solution mass, while the volume percent (%v/v) is the percentage of solute volume in solution volume, i.e.,

$$\%\mathrm{w/w} = \frac{\text{Mass of solute}}{\text{Mass of solution}} \times 100\%, \tag{5.1}$$

$$\%\mathrm{v/v} = \frac{\text{Volume of solute}}{\text{Volume of solution}} \times 100\%. \tag{5.2}$$

The mass/volume percent (%w/v) is defined as the percentage of solute mass in grams (g) in solution volume in milliliters (mL), i.e.,

$$\%\mathrm{w/v} = \frac{\text{Mass of solute (g)}}{\text{Volume of solution (mL)}} \times 100\%. \tag{5.3}$$

The mass/volume percent is not a relative unit as the numerator and denominator have different dimensions, but it is often used given that the mass of solute and the volume of solution are readily measured during solution preparation. For example, one can prepare 200 mL of 2% w/v alginate aqueous solution by dissolving 4 g of alginate salt powder in 200 mL of distilled water. Some aqueous solutions have a density close to 1 g/mL and, in such cases, the mass/volume percent approximately equals the mass percent.

Sometimes, the aforementioned concentration percentages are not suitable for describing the concentration of highly dilute solutions and those involving chemical reactions. For highly dilute solutions, one can use parts per million (ppm), parts per billion (ppb), and/or parts per trillion (ppt). For example, dissolving 1 µg of solute in 1 g of solution makes a 1 ppm solution. Similarly, units of ppb and ppt describe the concentration of solutions with 1 ng of solute in 1 g of solution and 1 pg of solute in 1 g of solution, respectively. For solutions involving chemical reactions, the units are commonly based on moles, such as mole fraction, molarity, and molality. One mole of a substance contains exactly $6.02214076 \times 10^{23}$ (Avogadro's number) particles, which may be atoms, molecules, ions, or electrons. For example, one mole of water contains $6.02214076 \times 10^{23}$ water molecules, with a total mass of about 18.015 g. The mole fraction of solute is the ratio of moles of solute to the total number of moles in solution. Multiplying the mole fraction by 100, one can describe the concentration as a mole percent. The molarity of solution is the number of moles of solute per liter of solution, while molality is the number of moles of solute per kilogram of solvent.

Biomaterial solutions can be prepared either manually or automatically. Generally, manual preparation is conducted for those materials that easily dissolve in water-based solutions, while automatic preparation devices, such as a magnetic stirring apparatus (Fig. 5.1a) or rocking machine (Fig. 5.1b), are used for materials that dissolve over an extended time period. Containers such as beakers (Fig. 5.1c), flasks (Fig. 5.1d), and wide-mouth bottles (Fig. 5.1e) are often used to prepare

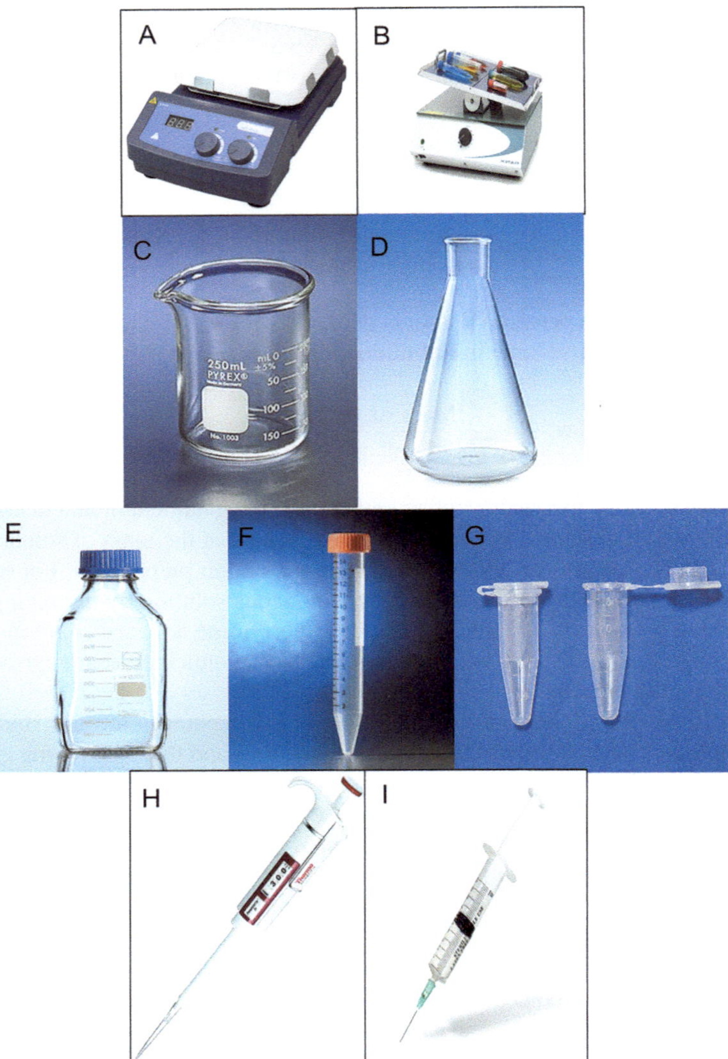

**Fig. 5.1** Commonly used devices and containers for biomaterial solution preparation. (**a**) magnetic stirring machine; (**b**) rocking machine; (**c**) beaker; (**d**) flask; (**e**) wide-mouth bottle; (**f**) centrifuge tube; (**g**) micro tube; (**h**) pipette; (**i**) syringe

biomaterial solutions in relatively larger volumes (over 50 mL), while containers such as centrifuge tubes (Fig. 5.1f) are normally employed for smaller solution amounts (less than 50 mL). If an even smaller amount of solution (less than 1 mL) is required, small capacity containers such as micro tubes (Fig. 5.1g) are more suitable, and homogeneous biomaterial solutions can be made with the help of pipettes (Fig. 5.1h) or syringes (Fig. 5.1i).

Prior to the preparation of scaffold solutions, the selected biomaterials may need to be sterilized to reduce or eliminate contamination by bacteria or other microorganisms. Techniques used to perform material sterilization include steam, dry heat, chemicals, and irradiation, all of which require abnormal conditions such as high pressure, elevated temperature, chemical toxicity, or high energy.

*Steam sterilization* or autoclaving uses pressurized steam at 121–134 °C to kill or deactivate microbes, bacterial spores, and viruses. While the intense heat released from high pressure steam can efficiently kill microbes by hydrolysis and coagulation of cellular proteins, high temperatures and/or long incubation times may also be required to kill some specific microorganisms.

*Dry heat* is used to kill microbes by oxidation of cellular components. This process requires more energy compared to hydrolysis, and thus dry heat sterilization requires temperatures higher than autoclaving temperatures. Dry heating can be provided by flaming (for metallic devices), incineration (especially for inoculation loops used in microbe cultures), or a hot air oven, which is suitable for dry materials such as powders, glassware, some metal devices, etc.

*Filtration* is another effective method to sterilize solutions. The filters should allow for the passage of the solution but stop, or filter out, larger microbes such as bacteria and viruses. Cellulose membrane filters are the most commonly used in microbial analysis and have varying pore sizes that are used for different purposes. Filters with a pore diameter of 0.2 μm, for example, can remove most bacteria; however, they may not remove viruses or phages owing to their smaller size, which can be less than 0.2 μm.

*Radiation sterilization* involves exposing materials to ultraviolet (UV) radiation, X-rays, and/or gamma rays. All of these are types of high-energy electromagnetic radiation that can profoundly damage DNA and, as such, are excellent sterilization tools. The main difference between the effectiveness of these rays is related to their penetration. UV has low penetration in air, and so is suitable for the sterilization of small areas (e.g., the interior of a biological safety cabinet) and is relatively safe. X-rays and gamma rays are more penetrating and thus more effective for large-scale sterilization. These rays are extremely dangerous, and special attention is required for their use. Large packages of medical devices can be sterilized using X-rays. Gamma rays are commonly used for the sterilization of food and disposable medical equipment (e.g., syringes, needles, cannulas, and IV sets).

*Chemical methods of sterilization* use harmful or toxic liquid and/or gas for sterilization of materials that may be damaged by heating. Ethanol is the most common disinfectant liquid, but isopropanol is a better solvent for fats and is often a superior option. Both of these solvents must be diluted to 60–90% in water to be effective. Although ethanol and isopropanol are effective at killing microbial cells, they do not affect spores. Due to their quick penetration into materials, gases are more effective than liquids for sterilization. A combination of ethylene oxide and carbon dioxide is commonly used for the sterilization of medical equipment sensitive to heat or moisture (e.g., catheters and stents). Adding carbon dioxide minimizes the chance of explosion. Ozone is another gas suitable for sterilization. Ozone is a powerful oxidant that destroys microorganisms and is compatible with a wide range of materials used for making reusable medical devices. Ozone is highly unstable and leaves no residues after sterilization. Vaporized hydrogen peroxide, formaldehyde steam, gaseous chlorine dioxide, and vaporized peracetic acid are some other examples of chemicals used for sterilization.

Sterilizing conditions may introduce toxic chemicals into the biomaterials and potentially result in molecular chain scission or molecular bonding within the biomaterial itself, which may change the material properties. For example, chemical agents such as ethylene oxide, which is toxic to cells, may be absorbed by a material during sterilization and then released later, negatively impacting cell viability. As such, when ethylene oxide and other similar chemicals are used for sterilization, an additional time period to release any residues is always suggested.

### 5.2.2   Solutions with Living Cells

To fabricate scaffolds with cell distributions that mimic native tissues/organs, scaffold solutions can be prepared with living cells (or bioinks) for bioprinting. As such, the prepared solutions must provide an aqueous environment that is both favorable for cell survival and suitable for printing. Currently, the most widely used biomaterials for cell bioprinting are hydrogels, as discussed in Chap. 3. Polymer solutions of hydrogels can provide a mild environment to ensure the vitality of mixed cells in the solutions, while the hydrogel formed after solidification contains a large amount of water and possesses similar properties to natural tissues. Hydrogel polymers are predominantly classified as either naturally derived or synthetic materials. Natural hydrogels are popular due to their inherent biocompatibility, while synthetic hydrogels have more uniform and predictable properties. The most commonly used hydrogels include alginate, chitosan, agarose, hyaluronic acid (HA), collagen, gelatin, fibrin, poly(ethylene glycol) (PEG), and poly(ethylene oxide) (PEO); decellularized matrix (dECM) components have also been developed as biomaterials for scaffold bioprinting with cells (Table 5.1) [3].

### 5.2.3   Solutions without Living Cells

Depending on the scaffold design and intended application, living cells may not be needed, and material solutions can be prepared without living cells for printing. For example, some scaffolds are produced as temporary structures to support the regeneration of damaged tissue by relying on the inherent recovery of the tissue itself. As is the case for solutions incorporating cells for bioprinting, solutions without cells also must be biocompatible and printable. As such, the biomaterials used for living cell printing can also be prepared as solutions for printing without cells.

One major challenge when building scaffolds using soft hydrogels is their limited mechanical properties. To address this challenge, some non-water-soluble polymers with high mechanical strength are often applied in extrusion bioprinting in combination with hydrogels to develop hybrid scaffolds [2, 4]. These materials are normally dissolved in solvent or melted inside the bioprinter by temperature control during the printing process, then extruded in a layer-by-layer pattern to form scaffold frameworks. Because the solvent used or elevated temperature environment can lead to serious problems with respect to cell and tissue survival, cells cannot be mixed in these materials. The most extensively used of such materials in extrusion bioprinting are polycaprolactone (PCL), polylactic acid (PLA), polyglycolic acid (PGA), and their copolymers.

## 5.3 Flow Behavior Characterization of Biomaterial Solutions

This section presents the classification of biomaterial solution flow behavior under shearing, along with common models used for its characterization.

### 5.3.1 Flow Behavior and Its Classification

**Newtonian Flow Behavior** Assume that a thin layer of fluid or solution is contained between two parallel plates (each with an area of $A$) located a distance ($L$) apart in the $Y$ direction, as shown in Fig. 5.2, with the top plate allowed to move in the $X$ direction under a force ($F$) while the bottom plate is fixed. In a steady-state condition, the force will be balanced by an equal and opposite internal frictional force in the fluid due to its viscosity. The frictional force per unit area is defined as the *shear stress*; to characterize the fluid movement under shearing, the *shear rate* is defined as the rate of change of velocity in the $Y$ direction (perpendicular to the shear stress). If the shear stress is directly proportional to the shear rate, this solution is known as a *Newtonian fluid*. The flow behavior of a Newtonian fluid is described by the linear relationship between the shear stress and shear rate, i.e.,

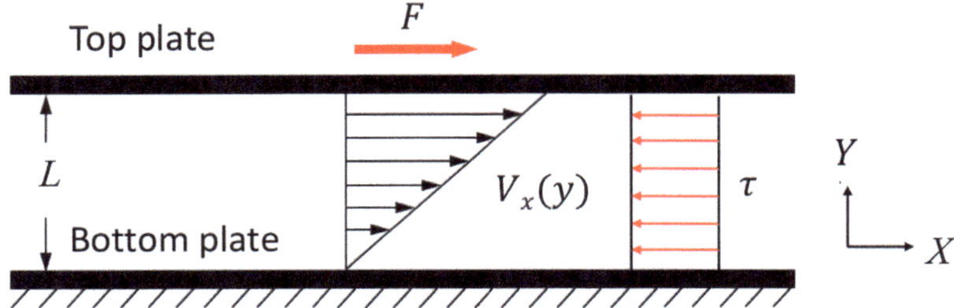


**Fig. 5.2** Schematic of flow behavior characterization under shearing

**Fig. 5.3** Flow curves of Newtonian, shear-thinning, viscoplastic, and shear-thickening solutions

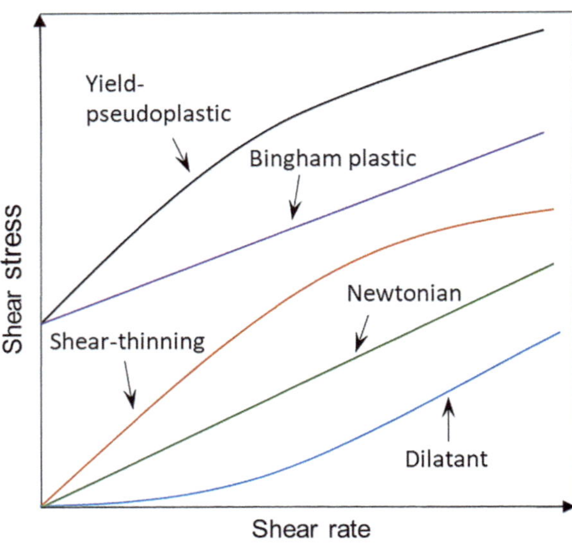


**Table 5.2** Newtonian fluids and their viscosities at room temperature

| Substance | Viscosity (mPa·s or cP) |
| --- | --- |
| Water | 1 |
| Ethyl alcohol | 1.20 |
| Ethylene glycol | 20 |
| Olive oil | 100 |
| Castor oil | 600 |
| Honey | 10,000 |

$$\tau = \mu \cdot \frac{dV_x(y)}{dy} = \mu \cdot \dot{\gamma}, \tag{5.4}$$

where $\tau$ is the shear stress (i.e., F/A) with a unit of Pa, $\dot{\gamma}$ is the shear rate with a unit of $s^{-1}$, and $\mu$ is the solution viscosity with a unit of Pa s.

The viscosity of a Newtonian fluid is constant regardless of the shear rate or shear stress and is only dependent on the solution itself or environmental conditions such as temperature. For a Newtonian fluid, the plot of shear stress vs. shear rate (also known as the flow curve) is a straight line that begins at the origin (Fig. 5.3), with a slope equal to the solution viscosity. Table 5.2 lists some Newtonian fluids in our daily lives with their viscosities at room temperature [5].

**Non-Newtonian Flow Behavior** For non-Newtonian solutions, the relationship between the shear stress and shear rate is not linear as their flow curves do not pass through the origin (Fig. 5.3). For a non-Newtonian solution, the viscosity is not a constant but varies depending on the shear stress, shear rate, and/or other conditions such as flow geometry or kinematic history of the solution. Generally, solutions with non-Newtonian flow behavior for scaffold printing can be divided into three groups:

- Solutions for which the shear stress at any point is determined only by the shear rate. These are known as time-independent solutions.
- Solutions for which the shear stress is not only dependent on shear rate but is also related to the duration of shearing and the kinematic history. These are called time-dependent solutions.
- Solutions that display both fluidic and elastic characteristics, showing partial elastic recovery after deformation. These are known as viscoelastic solutions.

### 5.3.1.1   Time-Independent Flow Behavior

Based on the relation between shear stress and shear rate for time-independent flow, solutions can be further classified as shear-thinning, viscoplastic, or shear-thickening.

Shear-thinning or pseudoplastic flow is the most common flow type for a time-independent non-Newtonian solution. The flow curve of a shear-thinning solution begins at the origin and increases in the shape of a concave curve. That is, an increasing shear rate leads to a smaller increase in shear stress than the proportional relation (Fig. 5.3). If the shear viscosity value at very low shear stress is $\mu_0$ and at infinite shear stress is $\mu_{inf}$, then the apparent viscosity of a shear-thinning fluid will decrease from $\mu_0$ to $\mu_{inf}$. Materials with shear-thinning properties are preferred for extrusion bioprinting because the decrease in material viscosity requests less force for printing. Thus, the material solutions are easily forced through the syringe and needle in the bioprinting process.

Viscoplastic fluid flow may not commence until a threshold value of stress, known as the yield stress $(\sigma_0)$, is exceeded. Such a material solution will deform elastically if the applied stress is less than the yield stress. Once the applied stress exceeds the yield stress, the fluid begins to flow and exhibits either a linear or non-linear flow curve that does not pass through the origin (Fig. 5.3). A solution with a linear flow curve is called a Bingham plastic fluid and is characterized by a constant viscosity. A solution with a non-linear flow curve after the yield point is known as yield-pseudoplastic and possesses partial characteristics of shear-thinning behavior.

Shear-thickening fluids display increasing viscosity as the shear rate increases (Fig. 5.3). This flow type was originally observed in concentrated suspensions such as gelatinized starch dispersions. The term dilatant is normally used to describe shear thickening, because these fluids expand slightly as the shear rate increases, leading to increased void space but insufficient liquid to fill it, increased friction between solid-solid contacts, and a viscosity increase. Dilatancy implies an increase in the volume of the solution but may be inappropriate for describing shear-thickening rheological behavior; thus, using the increased size of structural units under shearing to explain fluid shear-thickening behavior is more appropriate.

### 5.3.1.2   Time-Dependent Flow Behavior

In many cases, the apparent viscosity of a solution is not only determined by the shearing conditions and environment but also the time that the solution is exposed to shearing. For example, when a bentonite-water suspension is sheared under a constant rate, the apparent viscosity gradually decreases as the shearing period is extended. This phenomenon occurs due to the gradual breaking down of the internal structure of the material under external force. The broken structure of the material reduces its ability to prevent deformation due to external forces, resulting in the apparent viscosity decrease. As the structure breaks down, the rebuilding rate of internal structural linkages also increases. When the two rates become equivalent, a state of dynamic equilibrium is reached and a constant viscosity is maintained.

Generally, time-dependent flow behaviors include thixotropy and rheopexy. Thixotropy is when the apparent viscosity, or shear stress, decreases as the solution is sheared under a constant shear rate, as shown in Fig. 5.4. If the shear-thinning flow curve is plotted at a given shear rate from zero to a maximum value and then inversely back to zero, a hysteresis loop forms. Rheopexy is when the apparent viscosity increases with time of shearing, as shown in Fig. 5.4. The internal structures of a rheopectic fluid buildup during shearing and break down when the stress is removed. Hysteresis loops also form but with an inverted pattern compared to a thixotropic material solution. Only a few fluids express rheopexy under shearing, including some food solutions and aqueous gypsum pastes; such behavior is infrequent in scaffold bioprinting applications. With advances in biomaterials, more complex flow behaviors might become apparent, and such materials will require comprehensive analysis prior to their use in bioprinting.

**Fig. 5.4** Thixotropy (top) and rheopexy (bottom) flow curves

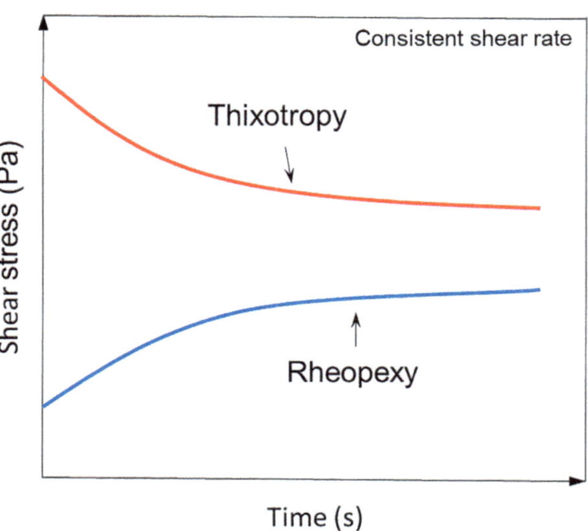

### 5.3.1.3   Viscoelastic Flow Behavior

A material solution normally displays viscosity as it is forced to flow under shearing. For an elastic material, the shear stress on the material is proportional to the strain; once the stress is removed, the material can quickly return to its original state. If a material solution shares the properties of both elasticity and viscosity, it is known as a viscoelastic material solution. Many materials such as melted polymers have both elastic and viscous characteristics as they are forced to flow or deform, exhibiting time-dependent strain and the ability to store and recover shear energy. Viscoelastic behavior is often seen in biomaterial solutions used for bioprinting.

## 5.3.2   Flow Behavior Models

Controlling the bioprinting process is essential to fabricating a scaffold as designed. This control, generally achieved by regulating printing conditions or parameters (including printing temperature, extrusion pressure, speed of the bioprinting head, and needle diameter), is basically determined by the scaffold design as well as the flow behaviors of the biomaterial solutions applied. It therefore becomes desirable to understand the flow patterns in the bioprinting process. The literature describes various flow behavior models that are available for this purpose. Some of these models are empirical attempts that give fitted relations for the shear stress and shear rate, while others have a theoretical basis from statistical mechanics. Commonly used models that describe time-independent behaviors are discussed in this section, including the power-law model, generalized power-law model, Carreau fluid model, Ellis fluid model, and Casson fluid model.

### 5.3.2.1   Power-Law Model

An expression of the power-law model is given by

$$\tau = K \cdot \dot{\gamma}^{n},\tag{5.5}$$

where $K$ is the consistency index with units of $Pa{\cdot}s^{n}$ and $n$ is the flow behavior index (dimensionless). Physically, $K$ is a measure of viscosity (higher $K$ means a more viscous the fluid), and $n$ is a measure of the degree of non-Newtonian behavior (greater departure of $n$ from unity means more pronounced non-Newtonian behavior). The value of these two parameters can be obtained from empirical curve fitting. For a Newtonian fluid as a special case, $n$ is equal to 1 and $K$ is equal to the viscosity of the fluid. If $n$ is less than 1, the fluid exhibits shear-thinning behavior; if $n$ is greater than 1, the fluid is shear-thickening. For non-Newtonian flow, the apparent viscosity is given by the power-law model:

$$\mu = \frac{\tau}{\dot{\gamma}} = K \cdot (\dot{\gamma})^{n-1}. \tag{5.6}$$

Note that the power-law model has been widely used to characterize the flow behavior of biomaterial solutions or bioink in bioprinting due to its simple formula for expressing the relation between the shear stress and shear rate; the empirically fitted curve can adequately represent and predict their flow behaviors.

### 5.3.2.2 Generalized Power-Law Model

The generalized power-law model is a modified power-law model that includes the yield stress of the biomaterial solution, $\tau_0$, as given by

$$\tau = \tau_0 + K \cdot \dot{\gamma}^n. \tag{5.7}$$

Tests using low shear rates are normally suggested to experimentally obtain the value of the yield stress. The recorded shear rates and related shear stress are used to estimate the yield stress at the point where the shear rate is equal to 0 [6, 7]. After determining the value of yield stress, the values of the other two parameters can be found following the same procedure used for developing a power-law model.

### 5.3.2.3 Carreau Fluid Model

To improve the accuracy of the power-law model in cases in which the flows have either low or high shear rates, Carreau developed a viscosity model based on considerations of the molecular network [8]. The model incorporates viscosities under both zero shear rate ($\mu_0$) and infinite shear rate ($\mu_{inf}$), as follows:

$$\frac{\mu - \mu_{inf}}{\mu_0 - \mu_{inf}} = \left(1 + (K \cdot \dot{\gamma})^2\right)^{\frac{n-1}{2}}, \tag{5.8}$$

where $n$ and $K$ are two curve-fitting parameters. This model can be used to describe shear-thinning flow within a wider range of shear rates compared to the power-law model. It is also more complex due to the addition of two other parameters.

### 5.3.2.4 Ellis Fluid Model

The Ellis fluid model is most suitable when the deviations from the power-law model are significant only at low shear rates. The Ellis model for simple shear conditions contains three constants:

$$\mu = \frac{\mu_0}{1 + \left(\tau/\tau_{\frac{1}{2}}\right)^{\alpha-1}}.$$

(5.9)

In this equation, $\alpha$ is a measured shear-thinning index, where larger values indicate a greater extent of shear-thinning. $\tau_{1/2}$ is the value of shear stress at the midpoint between zero and the maximum shear stress.

### 5.3.2.5   Casson Fluid Model

The Casson model was originally developed to characterize biological materials and is described as:

$$\tau^{\frac{1}{2}} = \tau_0^{\frac{1}{2}} + (K \cdot \gamma)^{\frac{1}{2}}.$$

(5.10)

In this equation, $\tau_0$ is the yield stress and $K$ is a constant, both of which can be determined from experimental data. This model has often been used to describe the steady shear stress and shear rate behavior of blood, yogurt, molten chocolate, and other foodstuffs and biological material solutions.

### 5.3.2.6   Time-Dependent Generalized Power-Law Model

Time-dependent biomaterial solutions or bioinks are said to be either thixotropic if the viscosity decreases with time, or rheopectic if the viscosity increases with time. One of the most promising models given in the literature to represent these behaviors is the time-dependent generalized power-law model. This model takes the form of

$$\tau = \tau_0 + K \cdot \dot{\gamma}^n \cdot t^m,$$

(5.11)

where $t$ is time and $m$ is the thixotropic index (dimensionless). Physically, $m$ is a measure of the degree of time dependency (the greater departure of $m$ from 0, the more pronounced the time dependency). Two notable drawbacks limit the application of this model. First, the equation cannot describe the equilibrium state of viscosity because, as time becomes very large, the second term on the right side of Eq. 5.11 reduces to 0 and the viscosity is only dependent upon the yield stress, which is obviously not true. The second is that this equation is not valid for the situation where, upon the removal of the shear stress, the viscosity gradually increases with time. This situation occurs during the time lapse between consecutive applications of shear stress due to the interruption of a printing process.

### 5.3.2.7   Structural-Theory Models

Models based on structural theory are one option to overcome the drawbacks of the time-dependent generalized power-law model. The general idea of structural theory is to use a structural parameter, $\lambda$, to account for the time-dependent effects on flow behavior. This parameter is associated with the internal structure of biomaterial solutions or bioink, which breaks down under shear and builds up at rest. Therefore, under the action of shear stress, the parameter continues to change until an equilibrium state is reached. One example of a structural-theory model takes the following form [6]:

$$\tau = \lambda(\tau_0 + K\dot{\gamma}^n) \tag{5.12}$$

$$\frac{d\lambda}{dt} = k_c(\lambda_i - \lambda) - k_d\lambda, \tag{5.13}$$

where $k_c$ and $k_d$ are rate constants associated with the build-up or creation and break-down or destruction of the internal structure of a biomaterial solution or bioink, respectively, and $\lambda_i$ is the initial value of the structural parameter. The time-dependent flow behavior can then be described by Eqs. 5.12 and 5.13 with the parameters $\tau_0$, $K$, $n$, $k_c$, $k_d$, and $\lambda_i$ evaluated from experimental data.

**Example 5.1**
A polymer solution is examined under both low and high shear rates at room temperature, with the data recorded in the following table. Select the model representative of the flow behavior of this solution and estimate the model parameters.

| Low shear rate (s$^{-1}$) | Shear stress (Pa) | High Shear rate (s$^{-1}$) | Shear stress (Pa) |
| --- | --- | --- | --- |
| 0.05 | 4.72 | 10 | 25.6 |
| 0.1 | 4.79 | 20 | 34.9 |
| 0.2 | 4.92 | 50 | 47.7 |
| 0.3 | 5.04 | 80 | 54.6 |
| 0.5 | 5.32 | 120 | 61.5 |
| 0.8 | 6.1 | 160 | 71.3 |
| 1.0 | 6.54 | 200 | 78.4 |

**Solution**

The data at the low shear rates in the above table indicates that a yield stress $\tau_0$ exists. Using the data at the low shear rates, the yield stress can be determined by linearly fitting the shear stress, as $\tau = \tau_0 + bx$. This provides a value for the yield stress of 4.54 Pa.

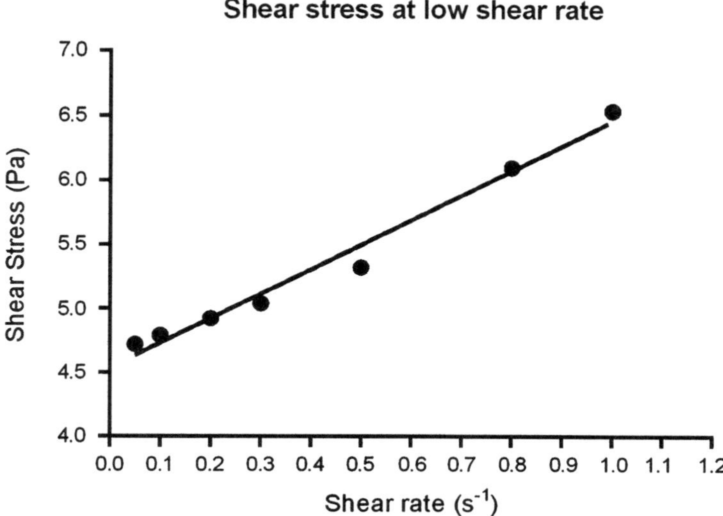



Under high shear rates, the shear stress exhibits non-linear behavior, which can be represented by the generalized power-law model. To determine the model parameters $K$ and $n$, rewriting Eq. 5.10 in a logarithmic form gives:

$$\log(\tau - \tau_0) = \log K + n \log \dot{\gamma}.$$

The values of $n$ and $K$ are then determined based on the linear regression of the measurement data of log $(\tau - \tau_0)$ versus log $(\dot{\gamma})$, where $\tau_0$ has a value of 4.54 Pa, as determined above. This process provides values of 8.713 for $K$ and 0.401 for $n$. Thus, the flow behavior of this polymer solution can be represented using the following Herschel-Bulkley fluid model:

$$\tau = 4.54 + 8.713 \times \dot{\gamma}^{0.401}.$$

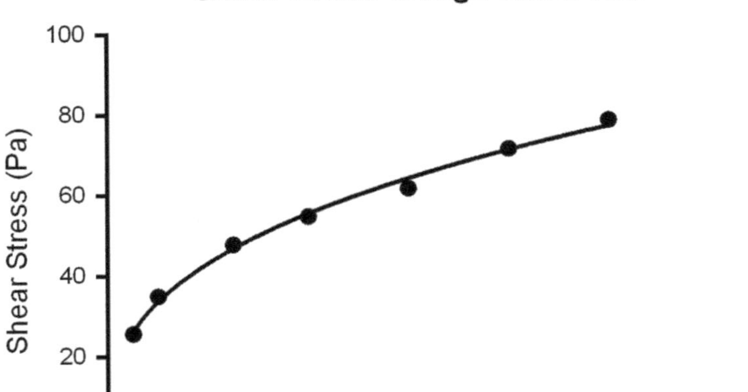



## 5.4   Techniques to Characterize Flow Behavior

This section discusses the commonly used techniques and/or equipment to measure and characterize the flow behavior of biomaterial solutions, including capillary rheometers, cone-and-plate rheometers, parallel plate rheometers, and oscillatory rheometers.

### 5.4.1  Capillary Rheometer

A capillary rheometer is basically a piston extruder with a capillary die at the end, as shown in Fig. 5.5. As the piston moves down, biomaterial solution loaded inside the reservoir is forced to extrude from the capillary. The shear stress in the capillary at the wall is related to the pressure drop along the capillary according to the following equation:

$$\tau = \frac{PD_c}{4L},\qquad(5.14)$$

where $P$ is the pressure drop in the capillary, and $D_c$ and $L$ are the diameter and length of the capillary, respectively. Assuming the piston diameter ($D_p$) is much larger than the capillary diameter (and neglecting entrance effects), then

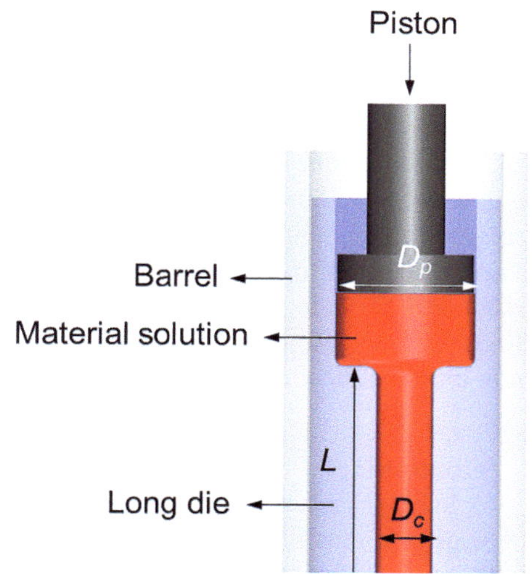

**Fig. 5.5** Schematic of a capillary rheometer



$$P = \frac{4F_p}{\pi D_p^{\,2}}.$$  (5.15)

From these equations, the shear stress at the capillary wall can be related to the force on the piston ($F_p$). Therefore, the shear stress can be determined by measuring the piston force. However, the capillary rheometer faces the problem of entrance effects when the ratio of capillary length to capillary diameter is small. To avoid this issue, a large ratio (higher than 20) is required. The shear rate at the capillary wall can be obtained from the volume flow rate, $Q$:

$$\dot{\gamma} = \frac{32Q}{\pi D_c^{\,3}},$$  (5.16)

and the flow rate can be determined from the area and velocity of the piston ($v_p$):

$$Q = v_p \frac{\pi}{4} D_p^{\,2}.$$  (5.17)

Thus, after measuring the velocity of the piston, the apparent viscosity can be calculated:

$$\mu = \frac{\tau}{\dot{\gamma}} = \frac{F_p D_c^{\,3}}{2\pi D_p^{\,4} v_P}.$$  (5.18)

This equation is valid for Newtonian fluids. If the fluid has non-Newtonian flow behavior and can be described by the power-law model (Eq. 5.3), the apparent shear rate at the wall is given by

$$\dot{\gamma} = \frac{(3n+1)8Q}{n\pi D_c{}^3},$$

(5.19)

where $n$ is the non-Newtonian fluid index. Thus, the flow behavior can be obtained.

The utilization of a capillary rheometer normally requires corrections for non-Newtonian fluids. The most important corrections are for the shear rate (referred to as the Rabinowitsch correction) and for the shear stress with respect to entrance effects (referred to as the Bagley correction) [9]. The advantages of the capillary rheometer include compatibility with high shear rates, ability to measure melt fractures, and ease of use. Its limitations include measuring corrections and locally non-uniform shear rates for the solutions examined.

**Example 5.2**
A hydrogel solution (density $\rho = 0.89$ g/mL) is being forced to flow through a capillary with a 500 µm inside diameter at a flow rate of 0.15 g/s. The flow is known to be laminar, and the power-law constants for the solution are $K = 25.5$ Pa·s$^n$ and $n = 0.36$. Evaluate the pressure drop over a 15 mm length ($L$) of a straight capillary.

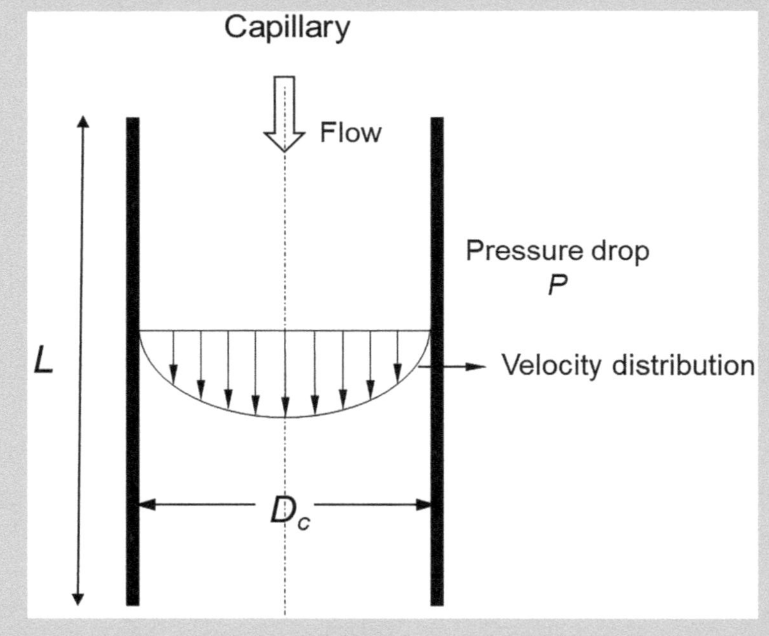


**Solution**
The volumetric flow rate $Q = 0.15/0.89 = 0.169$ mL/s $= 1.69 \times 10^{-7}$ mm³/s. According to Eq. 5.19, the apparent shear rate at the capillary wall is

$$\dot\gamma = \frac{(3n+1)8Q}{n\pi D_c^3} = \frac{(3 \times 0.36 + 1) \times 8 \times 1.69 \times 10^{-7}}{0.36 \times \pi \times \left(5 \times 10^{-4}\right)^3} = 19891.96\ s^{-1}.$$

Therefore, the shear stress at the capillary wall can be calculated using a power-law model (Eq. 5.5):

$$\tau = K \cdot \dot\gamma^n = 25.5 \times 19891.96^{0.36} = 899.63\ \text{Pa}.$$

Based on Eq. 5.14, the pressure drop can be calculated as

$$P = \frac{4L\tau}{D_c} = \frac{4 \times 0.015 \times 899.63}{5 \times 10^{-4}} = 107.96\ \text{kPa}.$$

### 5.4.2   Cone-and-Plate Rheometer

A schematic of a cone-and-plate rheometer is shown in Fig. 5.6. The test sample of a material solution is loaded into the gap between the cone and the plate of the rheometer and sheared under torsion. During testing, the cone is driven by a motor to rotate while the plate remains stationary. Let the angular velocity of the cone be denoted by $\omega$ and the cone angle by $\theta$. The shear rate within the material solution at the radius of $r$ is given by

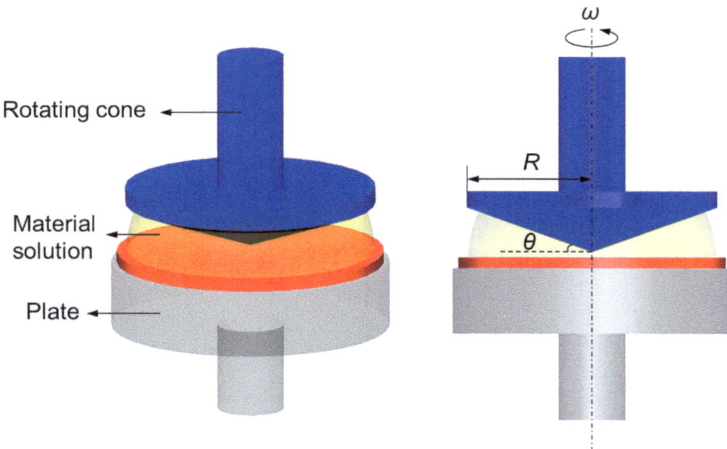


**Fig. 5.6** Schematic of a cone-and-plate rheometer

$$\dot{\gamma} = \frac{dv_s}{dy} = \frac{d(\omega \cdot r)}{d(r \cdot tan\theta)}, \tag{5.20}$$

where $v_s$ is the velocity at $r$, and $y$ is the distance from the stationary plate.

Typically, the cone is made very obtuse ($\theta < 4°$), and this small cone angle ensures the shear rate is approximately constant throughout the solution loaded in the gap, as $tan\theta \approx \theta$ in Eq. 5.20. Thus, the shear rate within the solution is given by

$$\dot{\gamma} = \frac{\omega}{\theta}. \tag{5.21}$$

The shear stress is related to the torque $T$ applied by the rotating cone by

$$\tau = \frac{3T}{2\pi R^3}. \tag{5.22}$$

The apparent viscosity can be obtained from the rheometer using

$$\mu = \frac{\tau}{\dot{\gamma}} = \frac{3T\theta}{2\pi R^3 \omega}. \tag{5.23}$$

Therefore, the flow behavior or apparent viscosity of a solution can be characterized from the torque ($T$) values recorded at varying speeds ($\omega$). For a non-Newtonian solution, apparent viscosity is not a constant but rather a function of shear rate ($\dot{\gamma}$). Also, for many cone-and-plate rheometers, the lowest shear rate that can be reached is normally greater than 1 s$^{-1}$, and therefore evaluating flow behavior under very low shear rates is challenging.

### 5.4.3   Parallel Plate Rheometer

A parallel plate rheometer is similar to a cone-and-plate rheometer but with an upper rotating flat plate (instead of a conical plate) and a lower stationary plate, as shown in Fig. 5.7. In contrast to a cone-and-plate geometry, the shear strain is proportional to the gap height, which can be varied to adjust the sensitivity of shear rate and thus the wall slip effect. Similar to a cone-and-plate rheometer, the shear stress and shear rate at $r$ for the parallel plate geometry are given as:

$$\tau = \frac{2T}{\pi r^3}, \tag{5.24}$$

$$\dot{\gamma} = \frac{\omega r}{h}, \tag{5.25}$$

where $h$ is the gap height between the two plates.

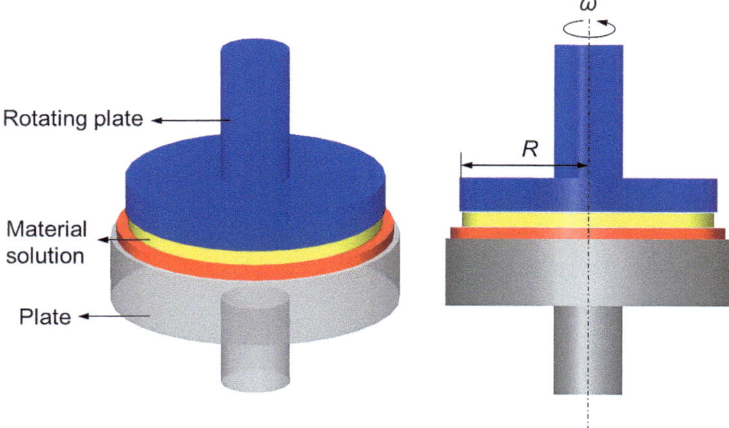


**Fig. 5.7**  Schematic of a parallel plate rheometer

The large gap sizes available can be used to overcome the limitations faced by the utilization of the cone-and-plate configuration, such as its sensitivity to eccentricities and misalignment. Loading and unloading fluids in the parallel plate rheometer are easier than for the cone-and-plate rheometer, particularly when evaluating fluids with high viscosity. Additionally, the parallel plate rheometer can more easily generate high shear rates by either increasing the angular speed of the upper steel plate or decreasing the gap size. When evaluating fluids in a wide gap, however, the shear rate is not constant throughout the loaded fluid. The strain reported is usually that measured at the outer rim, which provides a maximum value of the spatially varying strain within the gap.

## 5.4.4   Oscillatory Rheometer

The oscillatory shear technique has been widely used to characterize the viscoelastic behavior of biomaterials and/or their solutions. For this, an oscillatory rheometer is used to induce an oscillatory (e.g., sinusoidal) shear deformation in the sample and measure the resultant stress response, as shown in Fig. 5.8a, where the top plate is driven by a motor to impose a time-dependent shear deformation on the sample and the bottom plate remains stationary. Simultaneously, the time response of stress is characterized by measuring (via a load cell) the torque imposed on the bottom plate by the sample.

For a sinusoidal shear deformation, with a frequency $\omega$ and an amplitude $\gamma_0$,

$$\gamma(t) = \gamma_0 \, sin(\omega t). \tag{5.26}$$

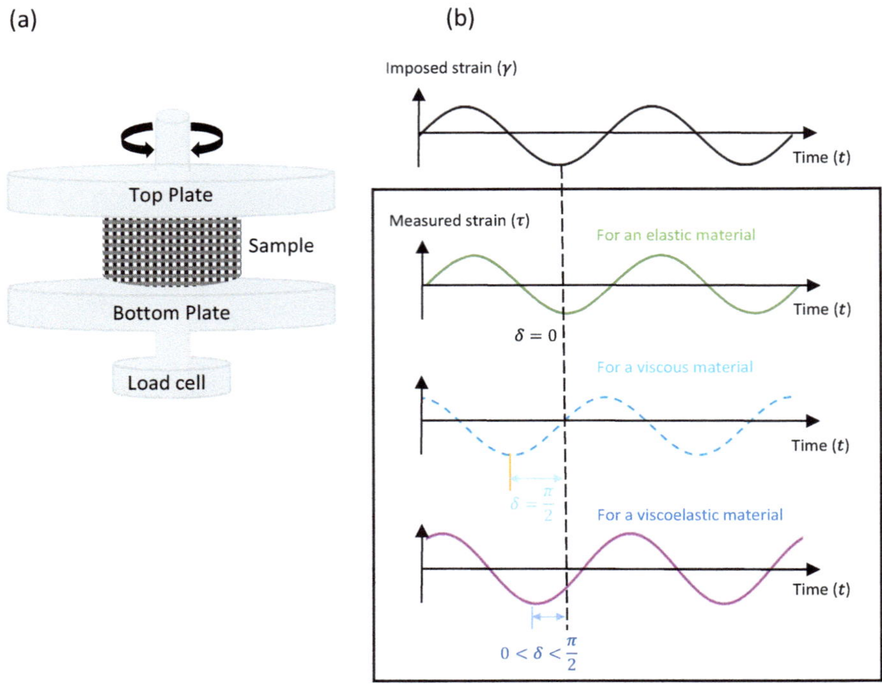



**Fig. 5.8** (**a**) Schematic representation of an oscillatory rheometer, with a sample placed between two plates, and (**b**) imposed sinusoidal strains and measured stresses of elastic, viscous, and viscoelastic materials, respectively

The shear rate can be determined by differentiating the shear strain with respect to time:

$$\dot{\gamma}(t) = \gamma_0 \omega \cos(\omega t) = \dot{\gamma}_0 \cos(\omega t). \tag{5.27}$$

Materials demonstrate linear viscoelastic behavior when the strain is sufficiently small. In the linear region, the behavior of a solution, such as the shear viscosity, is independent of the amplitude of the strain or strain rate. An oscillatory rheometer can be used to determine the linear viscoelastic properties of material solutions, and the measured shear stress can be written as

$$\tau(t) = \tau_0 \sin(\omega t + \delta), \tag{5.28}$$

where $\tau_0$ is the amplitude of the shear stress, and $\delta$ is the phase angle between the stress and strain and is termed the loss angle. For an elastic material, there is no phase shift between stress and strain, thus $\delta = 0$; for a pure viscous material, the loss angle reaches its maximum value, i.e., $\delta = \pi/2$; and for a viscoelastic material, the loss angle is in the range $0 < \delta < \pi/2$, as shown in Fig. 5.8b.

Expanding Eq. (5.28) gives

$$\tau(t) = \gamma_0[G' \sin(\omega t) + G'' \cos(\omega t)], \qquad (5.29)$$

where $G' = \frac{\tau_0}{\gamma_0} \cos(\delta)$ is termed the in-phase modulus or storage modulus, and $G'' = \frac{\tau_0}{\gamma_0} \sin(\delta)$ is termed the out-of-phase modulus or loss modulus, which respectively characterize the solid-like and fluid-like contributions to the measured shear stress. In the other words, the storage modulus ($G'$) represents the ability of a viscoelastic material to store energy in an elastic manner, while the loss modulus ($G''$) represents the energy dissipated in the viscous component of the material. The relationships among the storage modulus, loss modulus, and loss angle are illustrated in Fig. 5.9. For a given material or solution, the storage modulus, loss modulus, and loss angle, as measured by the oscillatory rheometer, are typically not constant but depend on the frequency ($\omega$) and amplitude ($\gamma_0$) of the sinusoidal stain applied. For such cases, frequency sweep tests (e.g., from 0.01 to 100 Hz at a constant strain) and strain sweep tests (e.g., from 0.01 to 100% at a constant frequency of 1 Hz) are performed to measure and characterize the storage modulus ($G'$), loss modulus ($G''$), and loss angle ($\delta$) [10].

In the context of crosslinking, $G'$ and $G''$ of a biomaterial solution can be obtained from oscillatory shearing, with the values often used to evaluate the crosslinking state of the biomaterial. Typically, the intersection point of $G'$ and $G''$ of a gelling solution for a range of temperatures is often used as the crosslinking temperature point [11]. This information is useful for deciding the temperature

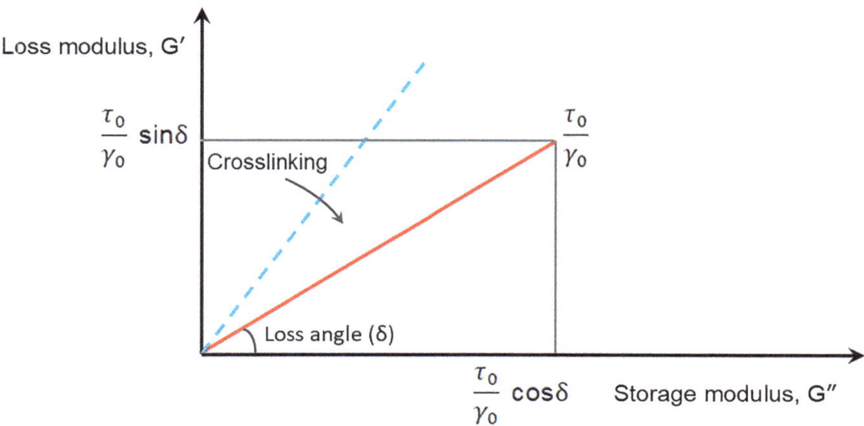


**Fig. 5.9** Illustration of the relationships among the storage modulus, loss modulus, and loss angle

required during the bioprinting process to regulate the gelation of a material with thermal crosslinking properties [10]. In the context of bioprinting, the loss angle of the bioink (or tan$\delta$) is known to directly impact its printability, thus providing a means to improve printability [10, 12].

## 5.5  Key Factors Related to Flow Behavior

Biomaterial solution flow behavior can be affected or regulated by many factors, including the properties of the biomaterial itself, material concentration, temperature, and addition of biological elements such as cells [13–17]. In this section, the influence of material concentration, temperature, and cell density on the flow behavior of biomaterial solutions is described, with alginate solution discussed as an example.

### *5.5.1  Influence of Material Concentration*

Material concentration indicates the amount of material dissolved in a given volume. A material solution with a higher concentration has more particles or molecules per unit volume of solution. The reduction in void space between molecules of the material as more is dissolved introduces more direct interactions between material molecules as the solutions are sheared, resulting in a more viscous solution [18].

For example, sodium alginate powder (alginic acid sodium salt from brown algae, A2033, Sigma-Aldrich, Canada) was thoroughly dissolved overnight in deionized water using a magnetic stirring machine to create a 0.2% w/v medium viscosity alginate solution. A 0.22-μm bottle-top filter (Thermo Scientific) was then used to sterilize the solution. The filtered solution was frozen and later lyophilized to obtain pure alginate powder, which was then dissolved again in calcium-free Dulbecco's modified Eagle's medium (DMEM) to form 1.0% and 2.0% w/v solutions [17].

The prepared alginate solutions were respectively loaded into the gap of a cone-and-plate rheometer (RVDV-III, Brookfield) according to the manufacturer's instructions. The cone was programmed to rotate at preset speeds that resulted in a predesigned range of shear rates (10 to 200 s$^{-1}$) during the shear process. The experiment was performed at room temperature. The shear stress at each shear rate point was recorded, and the flow curve plotted based on the recorded shear data is shown in Fig. 5.10. The two flow curves (for the 1.0% and 2.0% w/v solutions, respectively) show the alginate solutions exhibit shear-thinning, non-linear flow behavior. At the higher concentration, larger shear stress is measured at the same shear rate, demonstrating the increased concentration of alginate enhances the viscosity of the solution.

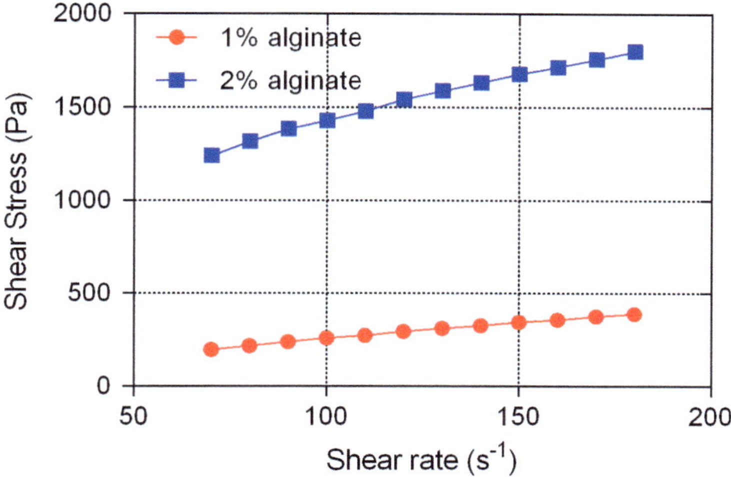


**Fig. 5.10**  Influence of alginate concentration on the flow behavior of alginate solutions [17]

## 5.5.2   *Influence of Temperature*

The molecules of a fluid are tightly bonded together by attractive intermolecular forces. These attractive forces are responsible for the viscosity of a fluid because they restrict the movement of individual molecules. As the temperature increases, the kinetic or thermal energy of the molecules increases, resulting in more mobile molecules and therefore reduced bonding energy, leading to a decrease in fluid viscosity.

To determine the influence of temperature on flow behavior, a 2% alginate solution was sheared at temperatures ranging from 15 to 40 °C [17]. The shear rate ranged from 10 to 150 s$^{-1}$ within this temperature range, with the resulting flow curves plotted in Fig. 5.11. The values of shear stress for the 2% alginate solution under the same shear rates decrease as the temperature increases, demonstrating the viscosity of the solution can be reduced by increasing the temperature. As the viscosity significantly affects the flow behavior of a solution, the regulation of printing temperature therefore becomes very important in the bioprinting process to control the flow behavior and flow rate of printed solutions for scaffold fabrication.

For cell-incorporated bioprinting, the cells are subjected to various stresses. To maintain their viability, the integrity of cell membranes must be maintained under various bioprinting temperatures. As more cells are injured as the bioprinting temperature increases, temperature control in scaffold bioprinting must consider not only the regulation of the flow behavior of the printed solution but also the conditions required to preserve cell viability [19].

Studies on the viscoelastic behavior of alginate solution [20] show increased temperature decreases the storage modulus ($G'$), loss modulus ($G''$), and viscosity. Figure 5.12 shows the effect of temperature on the storage and loss moduli of alginate solution.

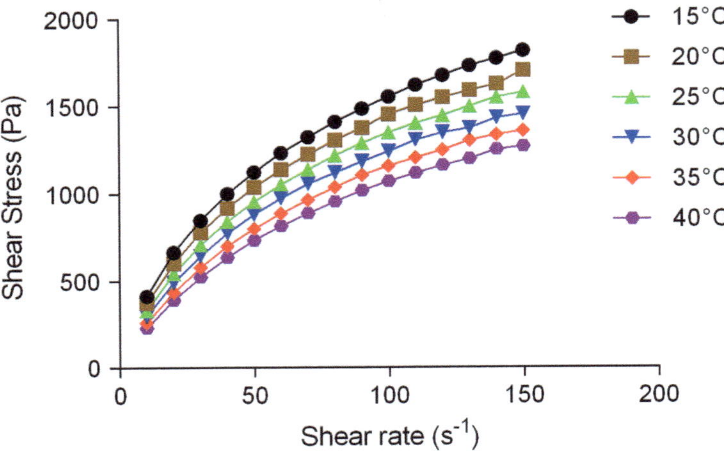

**Fig. 5.11**  Influence of temperature on the flow behavior of alginate solutions [17]

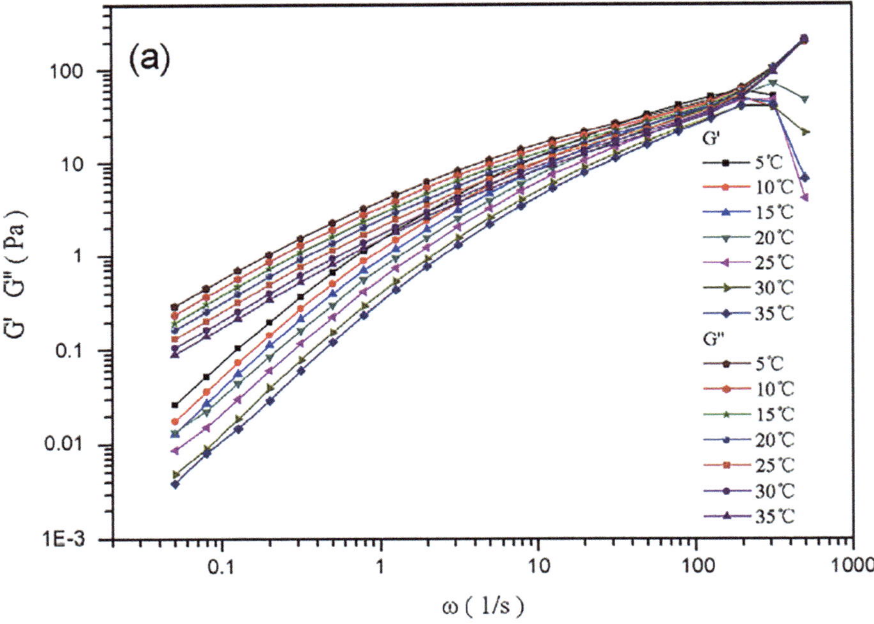

**Fig. 5.12**  Influence of temperature on storage and loss moduli (both are functions of frequency, $\omega$) of 2.5% (w/v) alginate solution [20]

### 5.5.3   *Influence of Cell Density*

Cells are the smallest structural and functional unit of an organism and consist of cytoplasm and a nucleus enclosed in a cell membrane. When cells are mixed into a biomaterial solution, they are 'non-soluble' microparticles suspended in the solution, changing the phase of uniformly dissolved biomaterial solution from a single phase to two phases. To determine the influence of cells on the flow behavior of a biomaterial solution, 1% alginate solutions with various cell densities (RSC96, a Schwann cell line, with cell densities of $5 \times 10^5$, $10^6$, $5 \times 10^6$, and $10^7$ cells/mL) were prepared [17]. The flow behavior of mixed solutions was then examined using a cone-and-plate rheometer at room temperature, with the flow curves plotted in Fig. 5.13.

The flow curves show the viscosity of the pure alginate solution is higher than the alginate solutions containing cells. As more cells are involved, solutions with lower viscosities can be achieved. Because cells contain cytoplasm, a fluid that is less viscous than the alginate solution, cell-cell and cell-biomaterial interactions likely act as lubricants within the alginate solution and thus reduce its viscosity [11, 21]. Also, the effect of cell density on the viscoelastic properties of bioinks has been reported, e.g., in a study on a gelatin/alginate bioink containing A549 cells [21].

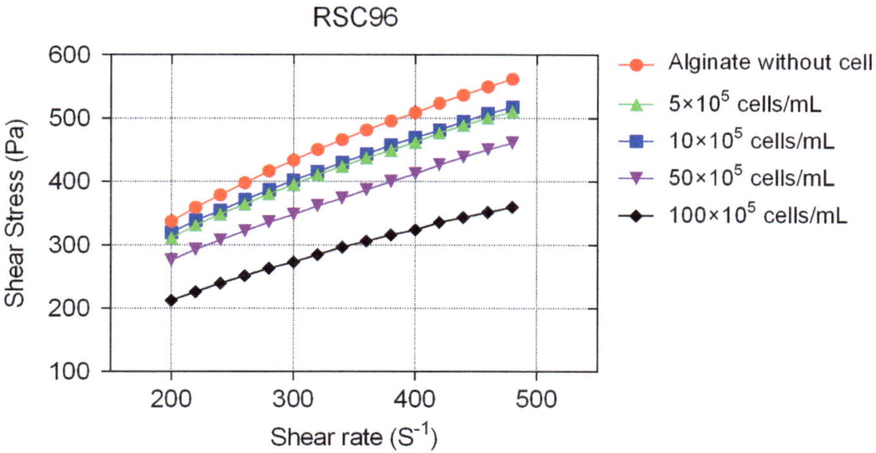


**Fig. 5.13**  Influence of cell density on the flow behavior of alginate solutions [13]

## 5.6    Summary

Preparing biomaterial solutions is a critical step in extrusion-based bioprinting. Biomaterial solutions should be prepared to provide a physiological environment along with sufficient mechanical support to protect incorporated cells, while having appropriate flow behavior for bioprinting. Biomaterials can be classified as water soluble or non-soluble. Among soluble materials, hydrogels and dECM have been widely used for preparing solutions with living cells due to their innate structural and compositional similarities to native ECM; however, the utilization of these materials can be limited by their weak mechanical properties. Non-soluble materials often display good mechanical support, but have drawbacks with respect to incorporating cells during the bioprinting process. Therefore, non-soluble materials are often used as frameworks to provide sufficient mechanical support for subsequently printed soft materials such as hydrogels. The concentration of a biomaterial solution represents the amount of biomaterials (as solutes) dissolved in a certain amount of solution. Prior to the preparation of scaffold solutions, the selected biomaterials typically need to be sterilized by means of steam, dry heat, chemical, and/or irradiation techniques.

The flow behavior of biomaterial solutions can be classified as Newtonian or non-Newtonian. If the relationship between shear stress and shear rate is linear, the solution is defined as Newtonian; if the relationship is nonlinear, the solution exhibits non-Newtonian behavior. Various models available in the literature can be used to represent the time-independent flow behavior, namely the power-law model, generalized power-law model, Carreau fluid model, Ellis fluid model, and Casson fluid model, each with advantages and disadvantages for representing the varying flow behaviors of biomaterial solutions. Time-dependent flow behaviors are either thixotropic if the viscosity decreases with time, or rheopectic if the viscosity increases with time; representative models include the time-dependent generalized power-law model and structural-theory models.

To experimentally measure the flow behavior of a biomaterial solution, commonly used techniques employ capillary rheometers, cone-and-plate rheometers, parallel plate rheometers, or oscillatory rheometers. The first three rheometers are typically used to measure and characterize the flow behavior described by the relationship between the shear stress and shear rate, while oscillatory rheometers are used to measure and characterize the viscoelastic behavior of biomaterial solutions.

The flow behavior of a biomaterial solution depends on the properties of the biomaterial used, material concentration, temperature, and cell density. For a given biomaterial (e.g., alginate), a higher material concentration enhances the viscosity of the solution, while higher temperatures and/or cell densities have the opposite effect. The flow behavior of a biomaterial solution is important with respect to its bioprinting to form scaffolds with a 3D structure.

**Problems**
1. What are the advantages and disadvantages of high- and low-viscosity biomaterial solutions in scaffold bioprinting?
2. What should be considered when preparing a biomaterial solution with cells for scaffold bioprinting?
3. To prepare two sterilized solutions of alginate (120 mL) and chitosan (10 mL), both at a concentration of 3% w/v, briefly explain the materials you would need and the protocol you would follow to prepare these solutions, including the containers, dissolving devices, and procedure.
4. Briefly explain the following concepts: shear stress, shear rate, yield stress, flow curve, Newtonian solutions, and non-Newtonian solutions.
5. If you realize that a biomaterial solution to be used in a bioprinting process is non-printable because of its low viscosity, what can you do to improve the situation? Give at least three options.
6. Why is the flow behavior of a biomaterial solution important in bioprinting? How are flow behaviors classified?
7. The following shear stress-shear rate data were obtained for a water-dissolved polymer solution at room temperature.

| Shear rate ($s^{-1}$) | Shear stress (Pa) | Shear rate ($s^{-1}$) | Shear stress (Pa) |
| --- | --- | --- | --- |
| 0.13 | 0.11 | 4.40 | 2.95 |
| 0.169 | 0.13 | 5.38 | 3.63 |
| 0.215 | 0.16 | 6.88 | 4.50 |
| 0.274 | 0.20 | 8.62 | 5.29 |
| 0.339 | 0.27 | 11.01 | 6.44 |
| 0.414 | 0.33 | 13.78 | 8.02 |
| 0.54 | 0.431 | 16.68 | 9.06 |
| 0.692 | 0.556 | 21.3 | 11.3 |
| 0.878 | 0.675 | 27.1 | 13.2 |
| 1.01 | 0.837 | 34.6 | 15.89 |
| 1.37 | 1.01 | 43.5 | 18.77 |
| 1.69 | 1.30 | 55.3 | 22.07 |
| 2.04 | 1.58 | 68.7 | 25.48 |
| 2.68 | 1.93 | 87.4 | 29.55 |
| 3.42 | 2.54 | 110.9 | 33.26 |

    (a) Can the power-law model fit these data over the entire range? What are the values of $K$ and $n$? Plot the curve.

    (b) Can the Ellis fluid model fit these data better than the power-law model? Evaluate the values of $\mu_0$, $\tau_{1/2}$, and $\alpha$.

8. Briefly explain one technique to characterize the flow behavior, along with an example reported in the literature.
9. Name and briefly explain one factor to affect the flow behavior of a biomaterial solution.

# References


1. A. Rajaram, D. Schreyer, D. Chen, Bioplotting alginate/hyaluronic acid hydrogel scaffolds with structural integrity and preserved Schwann cell viability. 3D Print Addit. Manuf. **1**, 194–203 (2014). https://doi.org/10.1089/3dp.2014.0006
2. Z. Izadifar, T. Chang, W. Kulyk, et al., Analyzing biological performance of 3D-printed, cell-impregnated hybrid constructs for cartilage tissue engineering. Tissue Eng. Part C Methods **22**, 173–188 (2016). https://doi.org/10.1089/ten.tec.2015.0307
3. F. Pati, J. Jang, D.H. Ha, et al., Printing three-dimensional tissue analogues with decellularized extracellular matrix bioink. Nat. Commun. **5**, 3935 (2014). https://doi.org/10.1038/ncomms4935
4. W. Schuurman, V. Khristov, M.W. Pot, et al., Bioprinting of hybrid tissue constructs with tailorable mechanical properties. Biofabrication **3**, 021001 (2011). https://doi.org/10.1088/1758-5082/3/2/021001
5. R.P. Chhabra, J.F. Richardson, *Non-Newtonian Flow in the Process Industries: Fundamentals and Engineering Applications* (Butterworth-Heinemann, Oxford, 1999)
6. X.B. Chen, Time-dependent rheological behavior of fluids for electronics packaging. J. Electron. Packag. **127**, 370–374 (2005). https://doi.org/10.1115/1.2056568
7. X. Chen, Dispensed-based bio-manufacturing scaffolds for tissue engineering applications. Int. J. Eng. Appl. **2**(1), 9–10 (2014)
8. P.J. Carreau, Rheological equations from molecular network theories. Trans. Soc. Rheol. **16**, 99–127 (1972). https://doi.org/10.1122/1.549276
9. J. Aho, S. Syrjälä, Determination of the entrance pressure drop in capillary rheometry using Bagley correction and zero-length capillary. Ann. Trans. Nord Rheol. Soc. **14**, 143–148 (2006)
10. N. Soltan, L. Ning, F. Mohabatpour, et al., Printability and cell viability in bioprinting alginate dialdehyde-gelatin scaffolds. ACS Biomater Sci. Eng. **5**, 2976–2987 (2019). https://doi.org/10.1021/acsbiomaterials.9b00167
11. T. Billiet, E. Gevaert, T. De Schryver, et al., The 3D printing of gelatin methacrylamide cell-laden tissue-engineered constructs with high cell viability. Biomaterials **35**, 49–62 (2014). https://doi.org/10.1016/j.biomaterials.2013.09.078
12. T. Gao, G.J. Gillispie, J.S. Copus, et al., Optimization of gelatin–alginate composite bioink printability using rheological parameters: A systematic approach. Biofabrication **10**, 034106 (2018). https://doi.org/10.1088/1758-5090/aacdc7
13. X.Y. Tian, M.G. Li, N. Cao, et al., Characterization of the flow behavior of alginate/hydroxy-apatite mixtures for tissue scaffold fabrication. Biofabrication **1**, 045005 (2009). https://doi.org/10.1088/1758-5082/1/4/045005
14. Z. Fu, S. Naghieh, C. Xu, et al., Printability in extrusion bioprinting. Biofabrication **13**, 033001 (2021). https://doi.org/10.1088/1758-5090/abe7ab
15. L. Ning, A. Guillemot, J. Zhao, et al., Influence of flow behavior of alginate–cell suspensions on cell viability and proliferation. Tissue Eng. Part C Methods **22**, 652–662 (2016). https://doi.org/10.1089/ten.tec.2016.0011
16. A. Zimmerling, Z. Yazdanpanah, D.M.L. Cooper, et al., 3D printing PCL/nHA bone scaffolds: Exploring the influence of material synthesis techniques. Biomater. Res. **25**, 3 (2021). https://doi.org/10.1186/s40824-021-00204-y
17. L. Ning, Y. Xu, X. Chen, D.J. Schreyer, Influence of mechanical properties of alginate-based substrates on the performance of Schwann cells in culture. J. Biomater. Sci. Polym. Ed. **27**, 898–915 (2016). https://doi.org/10.1080/09205063.2016.1170415
18. P. Vlasak, Z. Chara, Effect of particle size distribution and concentration on flow behavior of dense slurries. Part. Sci. Technol. **29**, 53–65 (2011). https://doi.org/10.1080/02726351.2010.508509

19. M.G. Li, X.Y. Tian, X. Chen, Temperature effect on the shear-induced cell damage in biofabrication. Artif. Organs **35**, 741–746 (2011). https://doi.org/10.1111/j.1525-1594.2010.01193.x
20. J. Ma, Y. Lin, X. Chen, et al., Flow behavior, thixotropy and dynamical viscoelasticity of sodium alginate aqueous solutions. Food Hydrocoll. **38**, 119–128 (2014). https://doi.org/10.1016/j.foodhyd.2013.11.016
21. B. Alberts, A. Johnson, J. Lewis, et al., *Molecular Biology of the Cell*, 4th edn. (Garland Science, New York, 2002)

# Chapter 6
# Bioprinting of Tissue Scaffolds

## 6.1 Introduction

Bioprinting refers to the process of printing the bioink or biomaterial solution with living cells and/or biomolecules to create or fabricate 3D constructs. Among the bioprinting techniques developed to date, extrusion bioprinting, which employs a pneumatic or other mechanism to extrude the bioink, has been widely used in the fabrication of scaffolds for tissue engineering. This chapter presents the basic principles of extrusion bioprinting, along with the influence of bioprinting process parameters on scaffold fabrication as well as the methods to characterize the structure of printed scaffolds. In the bioprinting process, the cells in the bioink are subject to process-induced mechanical forces that may rupture or damage them and thus reduce their viability and function after bioprinting. As such, this chapter also discusses these process-induced mechanical forces, their effects on cells, and techniques to characterize cell viability, which is followed by an introduction to several advanced bioprinting techniques.

## 6.2 Basics of Bioprinting Systems

Extrusion bioprinting systems, such as the one shown in Fig. 6.1, are configured to deposit or print continuous strands or fibers of bioinks to form 3D structures layer by layer. An extrusion-based bioprinting system typically resembles the configuration schematically shown in Fig. 6.2, which consists of a dispensing or printing head, a three-axis positioning system, and a printing stage, all controlled by a computer. The positioning system is used to move the printing head relative to the printing stage in the $X$, $Y$, and $Z$ directions, while the printing head deposits the bioink that is loaded in the syringe onto the printing stage. Temperature control is used to regulate the temperature of both the bioink and printing stage, typically in a range from


D. X. B. Chen, *Extrusion Bioprinting of Scaffolds for Tissue Engineering*,
https://doi.org/10.1007/978-3-031-72471-8_6

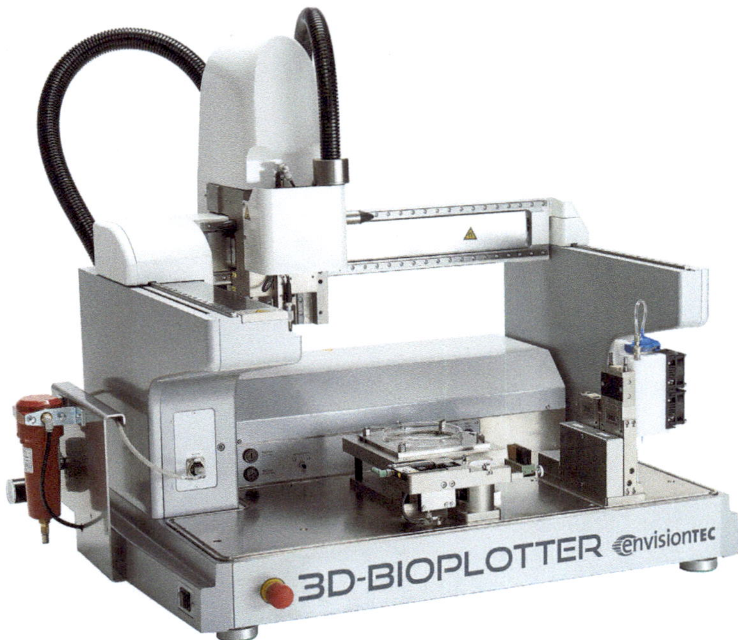



**Fig. 6.1** An extrusion-based bioprinting system for bioprinting

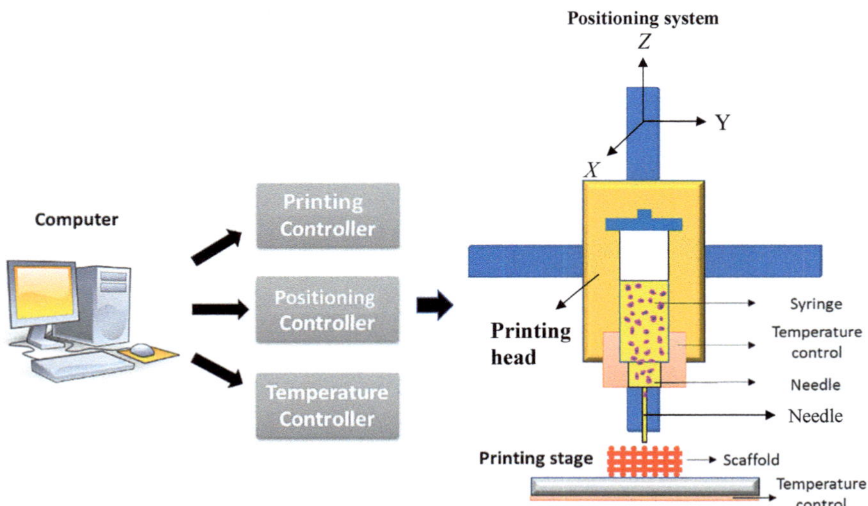



**Fig. 6.2** Schematic of a typical extrusion-based bioprinting system

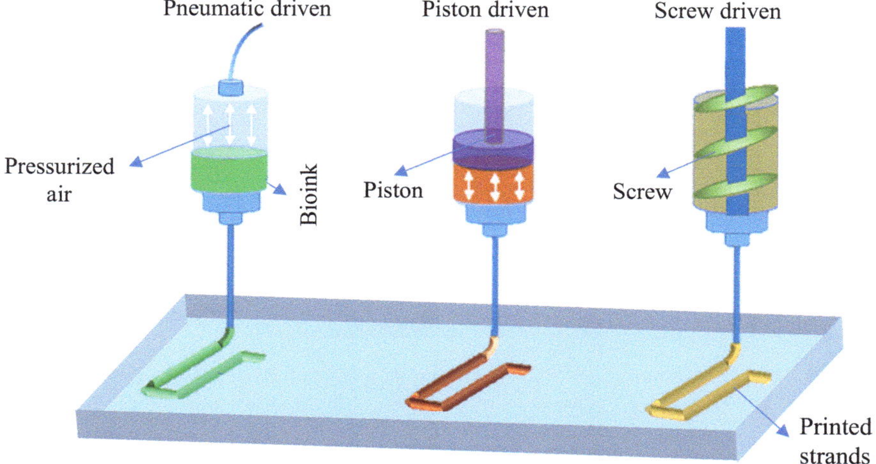


**Fig. 6.3** Schematic of pneumatic-, piston-, and screw-driven bioprinting systems

10 to 200 °C. The bioink loaded in the syringe is driven through the needle, which typically has either a cylindrical or tapered shape with a diameter ranging from 0.1 mm (or 100 μm) to 2 mm. The printed strand diameter and resolution are largely determined by the needle diameter, with higher resolution strands achievable by means of smaller needles.

The printing forces or mechanisms employed in extrusion bioprinting can be classified into three categories: pneumatic-, piston-, and screw-driven systems (Fig. 6.3). Pneumatic-driven bioprinting systems utilize pressurized air to drive the bioink out of the needle and, as such, deposition of the bioink is controlled by regulating the pressure of compressed air [1, 2]. Sterilization of the pressurized air and application of an air filter on the airway are recommended to minimize contamination of the bioink to be printed. Due to the advantages of simple operation and ease of maintenance, pneumatic-driven bioprinting has been widely used in scaffold fabrication. In position- or screw-driven bioprinting systems, bioinks inside the syringe are mechanically pushed out by a piston or a screw [3]. Both piston- and screw-driven systems can provide larger mechanical forces and also allow for more direct control over the flow of bioink compared to pneumatic-driven systems. The larger driving forces or pressures are of benefit when printing bioinks with higher viscosity, but more complex configurations, are required for piston- and screw-driven systems. In addition, large driving forces may rupture cell membranes and cause cell damage if living cells are incorporated in the printing process.

During the printing process, the movement of the printing head is regulated by the positioning controller based on the design of the scaffold to be printed, with the bioink driven from the needle onto the stage to form a scaffold. By altering the printing parameters, such as the speed of the printing head and the driving force, the bioink strands can be printed in varying sizes and/or patterns. Once one layer of

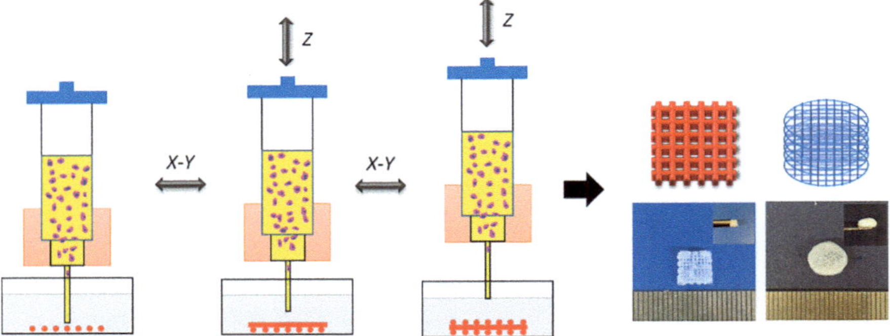

**Fig. 6.4** Schematic of multi-layer scaffold printing in a layer-by-layer pattern by controlling the positioning system in the *X-Y* plane and *Z* direction

strands is printed, the printing head is controlled to move up one step in the vertical or *Z* direction and then continue to print next layer. Through this process, scaffolds with 3D structures are printed on the stage, such as the ones shown in Fig. 6.4. In addition to the aforementioned process parameters (i.e., printing head speed and driving force), the scaffold structure can also be affected by other factors, including the temperature of both the bioink and printing stage.

## 6.3   Flow Rates in Bioprinting

During bioprinting, bioinks are printed to form scaffolds with a 3D structure by stacking the printed strands layer by layer and, as such, the characteristics of the printed strands affect the structure of the printed scaffold. The size of the printed strands is proportional to the flow rate of the bioink printed, i.e., the volume (or mass) of the bioink forced out of the needle per unit time. The flow rate can be affected by such factors as the process parameters (e.g., printing forces and temperature), structural parameters (e.g., needle geometry and size), and bioink flow behavior. One way to represent the flow rate in bioprinting is to develop empirical models from experimental data, but this typically requires exhaustive and time-consuming experiments. A more efficient way to represent the flow rate of the bioink being printed is based on physical laws, resulting in analytical models that form a basis to rigorously regulate the flow rate in the bioprinting process [4–7].

### *6.3.1   Pneumatic-Driven Bioprinting*

Pneumatic-driven bioprinting utilizes pressurized air to drive the bioink in the syringe to flow through and out of the needle, as shown in Fig. 6.5. To represent

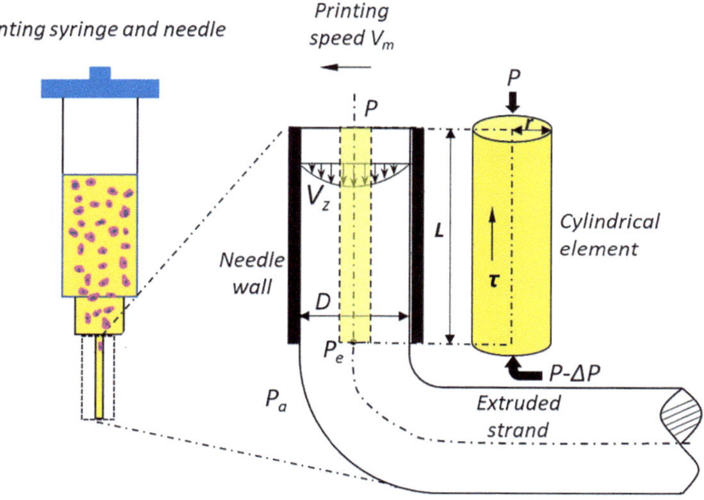


**Fig. 6.5** Schematic of pneumatic-driven bioprinting and bioink flow inside the needle

the flow rate of the bioink printed during the process, the following assumptions are made:

1. The air pressure in the syringe has reached a steady-state value, implying that the influence of air compressibility can be ignored.
2. The bioink being printed is incompressible and its flow behavior is time-independent.
3. The bioink flow inside the needle is laminar and fully developed.
4. There is no slip between the bioink and the needle wall.
5. The effect of gravity is neglected due to the fine needle size.
6. The pressure drop in the syringe can be ignored due to the fact that the internal diameter of the syringe is much greater than that of the needle.
7. The minor losses due to entrance effects and the minor losses in fittings can also be neglected given the fact that the flow velocity in the bioprinting process is small.

Under the above Assumption 6, the pressure drop ($\Delta P$) occurs only in the needle and is given by

$$\Delta P = P - P_e, \tag{6.1}$$

where $P$ is the pressure of compressed air applied in the pneumatic-driven bioprinting process and $P_e$ is the pressure in the bioink at the exit of the needle. Considering the effect of the bioink surface tension at the exit of the needle, $P_e$ is given by [4].

$$P_e = P_a + \frac{2\sigma}{D}, \tag{6.2}$$

where $P_a$ is the ambient air pressure (i.e., $1 \times 10^5$ Pa), $\sigma$ is the bioink surface tension in a unit of Pa/s, and $D$ is the internal diameter of the needle.

Now consider bioink in a cylindrical element with a radius, $r$, in the needle, as shown in Fig. 6.5. The equilibrium of forces acting on the bioink element gives

$$2\pi r L \tau = \Delta P \left( \pi r^2 \right), \tag{6.3}$$

where $\tau$ is the shear stress on the outside surface of the element and $L$ is the length of the element (same as the needle length). The term on the left side is the force acting on the outside surface of the element, resulting from the shear stress; the term on the right side is the force due to the pressure drop $\Delta P$. Rearranging Eq. (6.3) yields the shear stress on the element:

$$\tau = \frac{r \Delta P}{2L} \tag{6.4}$$

From Chapter 5, it is known that the shear rate $\dot{\gamma}$ inside the bioink describes the rate of velocity change in the radial direction, which can be expressed as.

$$\dot{\gamma} = \left( \frac{dV_z}{dr} \right), \tag{6.5}$$

where $V_z$ is the velocity along the axial direction at radius $r$.

From the bioink flow behavior discussed in Chap. 5, a relationship can be established between the shear stress and shear rate for a given bioink. If the bioink behavior is described by the Generalized Power Law equation, the relationship between the shear stress and shear rate can be rewritten as

$$\dot{\gamma} = \left( \frac{\tau - \tau_0}{K} \right)^{1/n} = \frac{1}{K^{1/n}} \left( \frac{r \Delta P}{2L} - \tau_0 \right)^{1/n} \tag{6.6}$$

Note that, due to the yield stress ($\tau_0$) of the bioink, the profile of the shear rate in the bioink is not continuous. As shown in Fig. 6.6, for the annular region with a radius of $r_0 = \frac{2L\tau_0}{\Delta P}$, the shear stress is not sufficiently large to overcome the yield stress ($\tau_0$), and so the shear rate is zero and the bioink flows as a plug; for the annular region from $r = r_0$ to $R$ (or at the needle wall), the shear stress in the bioink is greater than the yield stress ($\tau_0$), and so the shear rate is described by Eq. (6.6). The velocity profile of the bioink flowing in the needle can be obtained by the following integration:

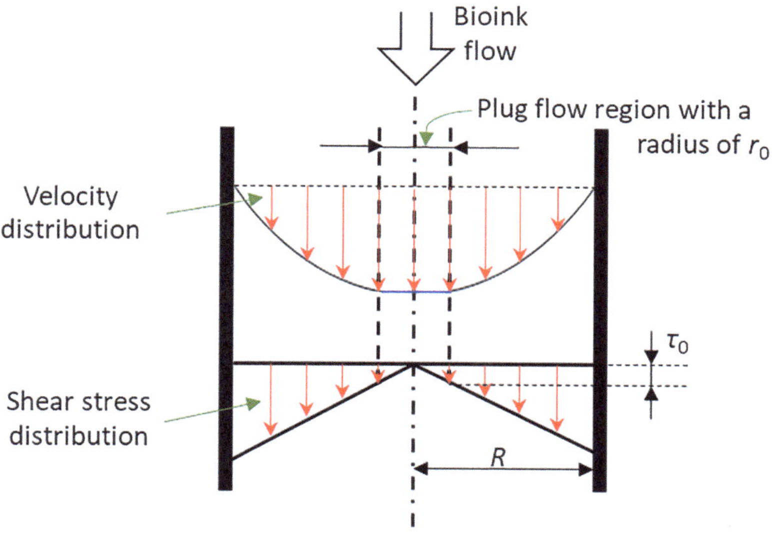


**Fig. 6.6**  Schematic velocity and shear stress distributions in the flow of a bioink with a yield stress ($\tau_0$)

$$V_z = \int_r^R \dot{\gamma}\,dr \tag{6.7}$$

Furthermore, for the annular region $r_0 < r < R$, the velocity gradually decreases from the constant plug velocity to zero at the needle wall and is given by.

$$V_{z1} = \frac{2Ln}{K^{1/n}\Delta P(n+1)}\left[\left(\frac{R\Delta P}{2L} - \tau_0\right)^{n+1/n} - \left(\frac{r\Delta P}{2L} - \tau_0\right)^{n+1/n}\right] \tag{6.8}$$

For the annular region $0 < r < r_0$, the bioink flows as a plug with a velocity that is obtained by substituting $r_0 = \frac{2L\tau_0}{\Delta P}$ into Eq. (6.8), i.e.,

$$V_{z2} = \frac{2Ln}{K^{1/n}\Delta P(n+1)}\left(\frac{R\Delta P}{2L} - \tau_0\right)^{n+1/n} \tag{6.9}$$

Equations (6.8) and (6.9) show that the velocity distribution of bioink inside the needle is not consistent, but rather a function of the needle radius. The velocity reaches a maximum at the needle center, and is zero at the needle wall. Based on the velocity, the volumetric flow rate $Q$ can be calculated by integrating the velocity over the needle cross-section:

$$
\begin{aligned}
Q &= \int_{r_0}^{R} 2\pi r V_{z1}\, dr + \int_{0}^{r_0} 2\pi r V_{z2}\, dr \\[2mm]
&= \frac{\pi R^3}{K^{\frac{1}{n}} \tau_w{}^3} (\tau_w - \tau_0)^{\frac{n+1}{n}} \left[ \frac{n}{3n+1}\tau_w{}^2 + \frac{2n^2}{(2n+1)(3n+1)}\tau_w \tau_0 \right. \\[2mm]
&\quad \left. + \frac{2n^3}{(n+1)(2n+1)(3n+1)} \tau_0{}^2 \right],
\end{aligned}
\tag{6.10}
$$

where $\tau_w$ is the shear stress at the needle wall and given by, from Eq. (6.4),

$$
\tau_w = \frac{R \Delta P}{2L}
\tag{6.11}
$$

If the bioink does not exhibit yield stress or $\tau_0 = 0$, the flow rate in Eq. (6.10) is reduced to.

$$
Q = \frac{n\pi R^{(3n+1)/n}}{(3n+1)(2KL)^{1/n}} (\Delta P)^{1/n}
\tag{6.12}
$$

If the bioink exhibits Newtonian flow, or $n = 1$, the flow rate in Eq. (6.12) is reduced to.

$$
Q = \frac{\pi R^4}{8KL} \Delta P
\tag{6.13}
$$

Equations (6.10), (6.12), and (6.13) illustrate that the flow rate of the bioink depends on its flow behavior (via $\tau_0$, $K$, and $n$), can be regulated by changing the printing process parameters (via $\Delta P$), and is affected by the structural parameters of the needle (via $R$ and $L$). With a known bioink flow behavior and parameter values, one can calculate the flow rate of the bioink printed in the process. These equations also provide a means to determine the printing pressure required for a given needle to achieve the desired flow rate for scaffold fabrication. Furthermore, by applying varying printing pressures and measuring the resulting flow rate, the flow behavior of the bioink can also be identified, as illustrated in Example 6.1, as an alternative method to using a rheometer [8].

**Example 6.1**

A bioink solution is prepared with a measured density of 1 g/mL (note that 1 g = 1000 mg, 1 mL = 1000 mm$^3$, or $1 \times 10^{-6}$ m$^3$, thus 1 g/mL = $1 \times 10^{9}$ mg/m$^3$). The bioink is printed at room temperature using an extrusion bioprinting system by means of a 200-μm diameter, 12-mm long cylindrical needle. During the printing process, the mass flow rate of the bioink at varying printing pressures is measured. The data obtained are listed in the table below.

| Pressure (kPa) | Mass flow rate (mg/s) |
| --- | --- |
| 20 | 0.000 |
| 40 | 0.003 |
| 60 | 0.016 |
| 80 | 0.055 |
| 100 | 0.137 |
| 120 | 0.289 |
| 140 | 0.554 |
| 160 | 0.974 |
| 180 | 1.643 |
| 200 | 2.528 |
| 220 | 3.610 |
| 240 | 5.418 |
| 260 | 7.394 |
| 280 | 9.990 |
| 300 | 13.420 |

1. If the bioink is a power-law fluid, identify the values of power-law parameters $K$ and $n$.
2. Find the volumetric flow rate and the flow velocity along the center of the needle for a printing pressure of 80 kPa.
3. Find the volumetric flow rate if a 330-μm diameter needle is used at a pressure of 100 kPa. Also determine the shear stress at the needle wall.

**Solution**

1. As the density $\rho$ is known, the mass flow rate $Q_m$ shown in the table can be converted into volumetric flow rate $Q$ (mm$^3$/s). From Eq. (6.12), we have.

$$Q_m = Q \cdot \rho = \rho \frac{n\pi R^{(3n+1)/n}}{(3n+1)(2KL)^{1/n}} (\Delta P)^{1/n}$$

Note that in the above equation, the units are

$$\frac{\text{mg}}{\text{s}} = \frac{\text{m}^3}{\text{s}} \cdot \frac{\text{mg}}{\text{m}^3} = \frac{\text{mg}}{\text{m}^3} \cdot \frac{\text{m}^{(3n+1)/n}}{(\text{Pa} \cdot \text{s}^n \cdot \text{m})^{1/n}} (\text{Pa})^{1/n}$$

$K$ and $n$ can be estimated using curve fitting from the data given in the above table, as shown in the following figure. This curve fitting gives $K = 103.45$ Pa·s$^n$ and $n = 0.2414$.

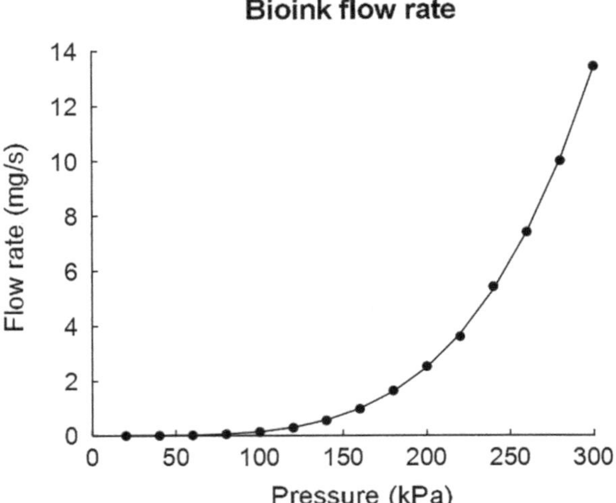

**Bioink flow rate**

2. For $\Delta P = 80$ kPa, $R = 100$ μm $= 100 \times 10^{-6}$ m, $L = 12$ mm $= 12 \times 10^{-3}$ m, along with $K = 103.45$ Pa·s$^n$ and $n = 0.2414$ identified above, the volumetric flow rate is evaluated using Eq. (6.12):

$$Q = \frac{n\pi R^{\frac{3n+1}{n}}}{(3n+1)(2KL)^{\frac{1}{n}}} (\Delta P)^{\frac{1}{n}}$$

$$= \frac{0.2414\pi \left(100 \times 10^{-6}\right)^{\frac{3 \times 0.2414 + 1}{0.2414}}}{(3 \times 0.2414 + 1)\left(2 \times 103.45 \times 12 \times 10^{-3}\right)^{\frac{1}{0.2414}}} \left(80 \times 10^3\right)^{\frac{1}{0.2414}}$$

$$= 5.60 \times 10^{-11} \frac{\text{m}^3}{\text{s}} \text{ or } 5.60 \times 10^{-5} \frac{\text{mL}^3}{\text{s}}$$

The velocity of the bioink along the centerline of the needle is evaluated from Eq. (6.8) with $r = 0$ and $\tau_0 = 0$:

$$V_z = \left(\frac{n}{n+1}\right)\left(\frac{\Delta P R}{2KL}\right)^{\frac{1}{n}} \cdot R = \left(\frac{0.2414}{0.2414 + 1}\right)\left(\frac{80 \times 10^3 \cdot 100 \times 10^{-6}}{2 \cdot 103.45 \cdot 12 \times 10^{-3}}\right)^{\frac{1}{n}} \cdot 100 \times 10^{-6}$$

$$= 2.48 \times 10^{-3} \text{ m/s}$$

3. If the diameter of the needle increases from 200 to 330 μm (or the radius increases from 100 to 165 μm), the volumetric flow rate at a pressure of 100 kPa is, from Eq. (6.12),

$$Q = \frac{0.2414\pi\left(165 \times 10^{-6}\right)^{\frac{3 \times 0.2414 + 1}{0.2414}}}{(3 \times 0.2414 + 1)\left(2 \times 103.45 \times 12 \times 10^{-3}\right)^{\frac{1}{0.2414}}}\left(100 \times 10^3\right)^{\frac{1}{0.2414}}$$

$$= 5.05 \times 10^{-9}\ \frac{m^3}{s}\quad \text{or } 5.05 \times 10^{-3}\ \frac{mL^3}{s}$$

The shear stress at the needle wall is, from Eq. (6.11),

$$\tau_w = \frac{R\Delta P}{2L} = \frac{165 \times 10^{-6} \times 100 \times 10^3}{2 \times 12 \times 10^{-3}} = 687.5\ \text{Pa.}$$

## 6.3.2   Screw-Driven Bioprinting

Figure 6.7a is a schematic of screw-driven bioprinting. The bioink is stored in a reservoir. Under the action of pressurized air, the bioink in the reservoir is fed, through a feed tube, to the chamber in a cartridge, where a screw, driven by a DC motor, rotates and forces the bioink down the chamber and eventually out of the needle. One of the distinct features of this approach is the ability to achieve high flow rates for fluids with a wide range of viscosities [9].

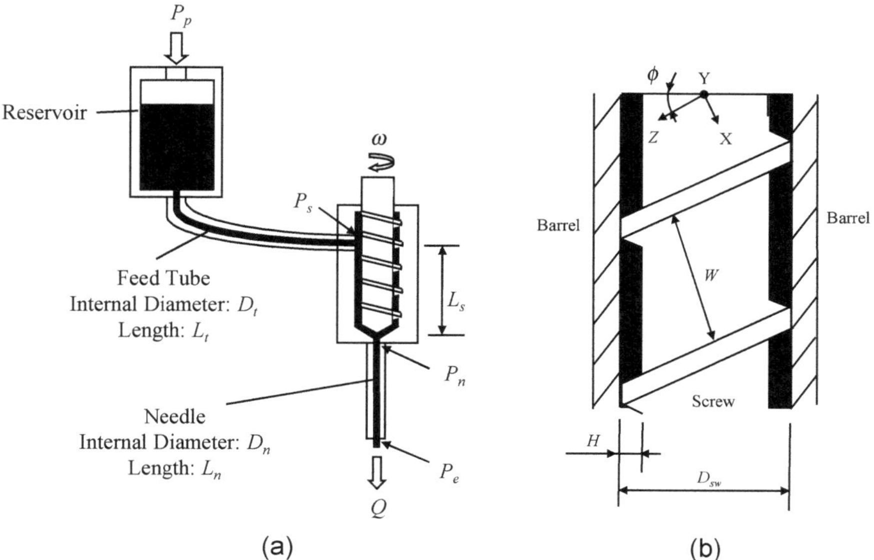


**Fig. 6.7**  Schematic of screw-driven bioprinting (**a**) and bioink flow inside the screw channel (**b**)

In Fig. 6.7a, the pressures of interest are denoted as follows: $P_p$ is the applied air pressure, $P_s$ is the fluid pressure at the entrance to the screw channel, $P_n$ is the fluid pressure at the needle entrance, and $P_e$ is the fluid pressure at the needle exit (given by Eq. (6.2)). $Q$ denotes the flow rate of the bioink being printed, which is the same as that in the feed tube, screw channel, or needle if the compressibility of the bioink is neglected. Figure 6.7b shows the flow in the screw channel, in which $D_{sw}$ denotes the outer diameter of the screw, $\phi$ is the helix angle of the screw, and $W$ and $H$ are the channel width and depth, respectively.

It is assumed that the internal surfaces of the screw root and barrel can be approximated to flat plates, given that the screw depth is much smaller than the barrel internal diameter in the bioprinting process. It is also assumed that (1) the flow of the bioink is steady and laminar and the behavior is time-independent; (2) there is no slip between the fluid and the barrel internal surface; (3) the leakage flow through the clearance between the flight and barrel is ignored; and (4) the flow is two dimensional in the down-channel and cross-channel directions. For simplicity in the following discussion, we further assume that the bioink exhibits Newtonian flow, or $n = 1$, so the flow rate in the screw channel is given by [9].

$$Q = c_{21}\omega - c_{22}(P_n - P_s) \tag{6.14}$$

where $c_{21}$ and $c_{22}$ are coefficients given by

$$c_{21} = \frac{WHD_{sw}\cos\varphi}{4}, \quad c_{22} = \frac{WH^3\sin\varphi}{12KL_s}$$

The bioink flow in both the feed tube and the needle is pressure driven. From Eq. (6.13), these flow rates are respectively given by

$$Q = c_1\left(P_p - P_s\right)^{1/n} \tag{6.15}$$

$$Q = c_3\left(P_n - P_e\right)^{1/n} \tag{6.16}$$

where the coefficients $c_1$ and $c_3$ are given as follows:

$$c_1 = \frac{\pi D_t^4}{128KL_t}, \quad c_3 = \frac{\pi D_n^4}{128KL_n}$$

The above equations show the flow rate is related to the pressure, either $P_s$ or $P_n$, both of which are unknown. By eliminating $P_s$ and $P_n$ from Eqs. (6.14), (6.15), and (6.16), the following relationship is obtained:

$$Q = C_\omega\omega + C_P(P_P - P_e) \tag{6.17}$$

where $C_\omega = \frac{c_1 c_3 c_{21}}{c_1 c_3 + c_1 c_{22} + + c_3 c_{22}}$ and $C_P = \frac{c_1 c_3 c_{22}}{c_1 c_3 + c_1 c_{22} + + c_3 c_{22}}$.

Equation (6.17) shows that the flow rate in screw-driven bioprinting involves two terms. The first term on the right-hand side of the equation is the rate of drag flow due to the screw rotation, which is proportional to the screw speed; the second term is the rate of pressure-driven flow, which is proportional to the pressure drop. Eq. (6.17) provides a means to control the flow rate of the bioink in screw-driven bioprinting by regulating the screw speed and/or printing pressure applied to the reservoir. The flow rate is also affected by the screw geometry ($D_{sw}$, $\phi$, $W$, and $H$) and flow behavior ($K$).

### *6.3.3   Piston-Driven Bioprinting*

Figure 6.8 is a schematic of piston-driven bioprinting, where the linear movement of the motor-driven piston is employed to drive fluid out of the needle. Under the assumption that the bioink is incompressible, the flow rate of the bioink is given by [10]:

$$Q = \dot{X} A_P \tag{6.18}$$

where $\dot{X}$ is the speed of piston movement and $A_p$ is the cross-sectional area of the piston. Eq. (6.18) indicates that the flow rate of the bioink is only dependent on the piston movement and not on the properties of the bioink being printed. As such,

**Fig. 6.8** Schematic of piston-driven bioprinting

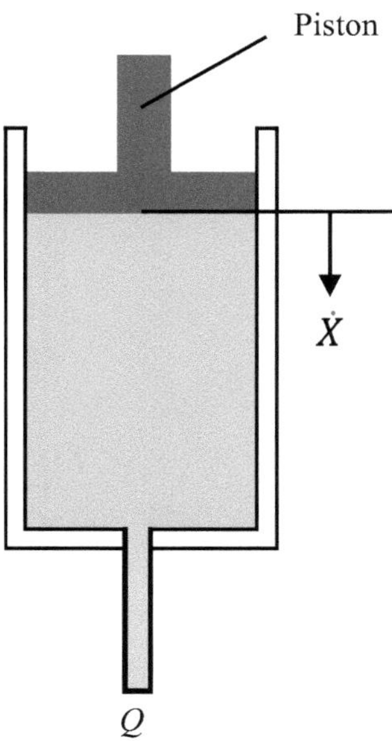

piston-driven bioprinting can provide the best means to regulate the flow rate of bioink compared to pneumatic- and screw-driven bioprinting.


### *6.3.4   Effect of Needle Geometry*

The needle geometry affects the flow rate of bioink printed in the bioprinting process. This section discusses this effect by examining the pressure-drive flow in needles with different geometries (focusing on cylindrical vs. tapered needles).

Figure 6.9 shows bioprinting processes employing cylindrical and tapered needles, and Fig. 6.10 shows some examples of these needles or tips. In Fig. 6.9, $\Delta P$ denotes the pressure drop in the needle; $D_c$ and $L_c$ are the internal diameter and length of the cylindrical needle, respectively; $D_i$ and $D_o$ are the diameters at the entrance and exit of the tapered needle, respectively; $\theta_0$ is the half cone angle; and $L_t$ is the length of the tapered needle. The half cone angle can be expressed as a function of length $L_t$ and diameters $D_i$ and $D_o$, i.e.,

$$\theta_0 = \arctan\left(\frac{D_i - D_o}{2L_t}\right) \tag{6.19}$$

For the pressure-driven flow shown in Fig. 6.9, the same assumptions are made as in Sect. 6.3.1. For simplicity, we consider a bioink with flow behavior described by the power-law equation with $\tau_0 = 0$. The flow rate of bioink printed from the cylindrical needle is given by Eq. (6.12), while that from the tapered needle is given by [7].

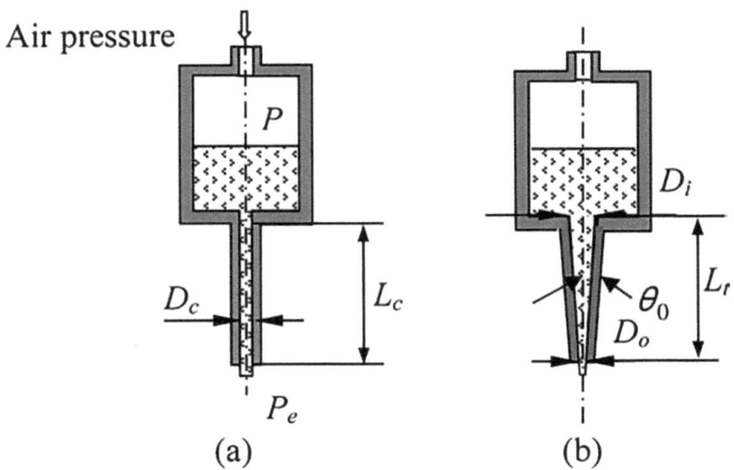



**Fig. 6.9** Bioprinting processes using (**a**) a cylindrical needle and (**b**) a tapered needle

| Gauge | Color | | ID | | OD | |
|---|---|---|---|---|---|---|
| | | | mm | inch | mm | inch |
| 14 | | Olive | 1.54 | 0.060 | 1.83 | 0.072 |
| 15 | | Amber | 1.36 | 0.053 | 1.65 | 0.065 |
| 18 | | Green | 0.84 | 0.033 | 1.27 | 0.050 |
| 20 | | Pink | 0.61 | 0.024 | 0.91 | 0.036 |
| 21 | | Purple | 0.51 | 0.020 | 0.82 | 0.032 |
| 22 | | Blue | 0.41 | 0.016 | 0.72 | 0.028 |
| 23 | | Orange | 0.33 | 0.013 | 0.65 | 0.025 |
| 25 | | Red | 0.25 | 0.010 | 0.52 | 0.020 |
| 27 | | Clear | 0.20 | 0.008 | 0.42 | 0.016 |
| 30 | | Lavender | 0.15 | 0.006 | 0.31 | 0.012 |
| 30 | | Black | 0.15 | 0.006 | 0.31 | 0.012 |
| 32 | | Yellow | 0.10 | 0.004 | 0.24 | 0.009 |

**SMOOTHFLOW TAPERED TIPS**

| Gauge | Color | | ID | |
|---|---|---|---|---|
| | | | mm | inch |
| 14 | | Olive | 1.60 | 0.063 |
| 16 | | Grey | 1.19 | 0.047 |
| 18 | | Green | 0.84 | 0.033 |
| 18 | | Black | 0.84 | 0.033 |
| 20 | | Pink | 0.58 | 0.023 |
| 20 | | Black | 0.58 | 0.023 |
| 22 | | Blue | 0.41 | 0.016 |
| 25 | | Red | 0.25 | 0.010 |
| 25 | | Black | 0.25 | 0.010 |
| 27 | | Clear | 0.20 | 0.008 |
| 27 | | Black | 0.20 | 0.008 |

**Fig. 6.10** Needles or tips used in bioprinting (ID, internal diameter; OD, outer diameter; reprinted with permission from EFD)

$$Q = \frac{\pi D_i^3 D_o^3}{32} \left[ \frac{3n\Delta P \tan\theta_0}{2K\left(D_i^{3n} - D_o^{3n}\right)} \right]^{\frac{1}{n}} \tag{6.20}$$

The above equation indicates that the flow rate of the bioink printed through a tapered needle depends on the pressure drop and can be affected by both flow behavior and needle geometry parameters.

**Case Study 6.1** In this case study [7], the bioink was prepared from alginate solution with cells (Schwann cells or 3 T3 fibroblasts) for bioprinting, and flow rates of the bioink printed from both cylindrical and tapered needles of the same exit diameter (i.e., $D_c = D_o$) in Fig. 6.9) were measured and compared under varying bioprinting conditions. The length of all needles was fixed at 20 mm and the flow behavior of the bioink was characterized and described by the power-law model with parameter values of $K = 10.82$ Pa·s$^n$ and $n = 0.54$. Figure 6.11a shows the

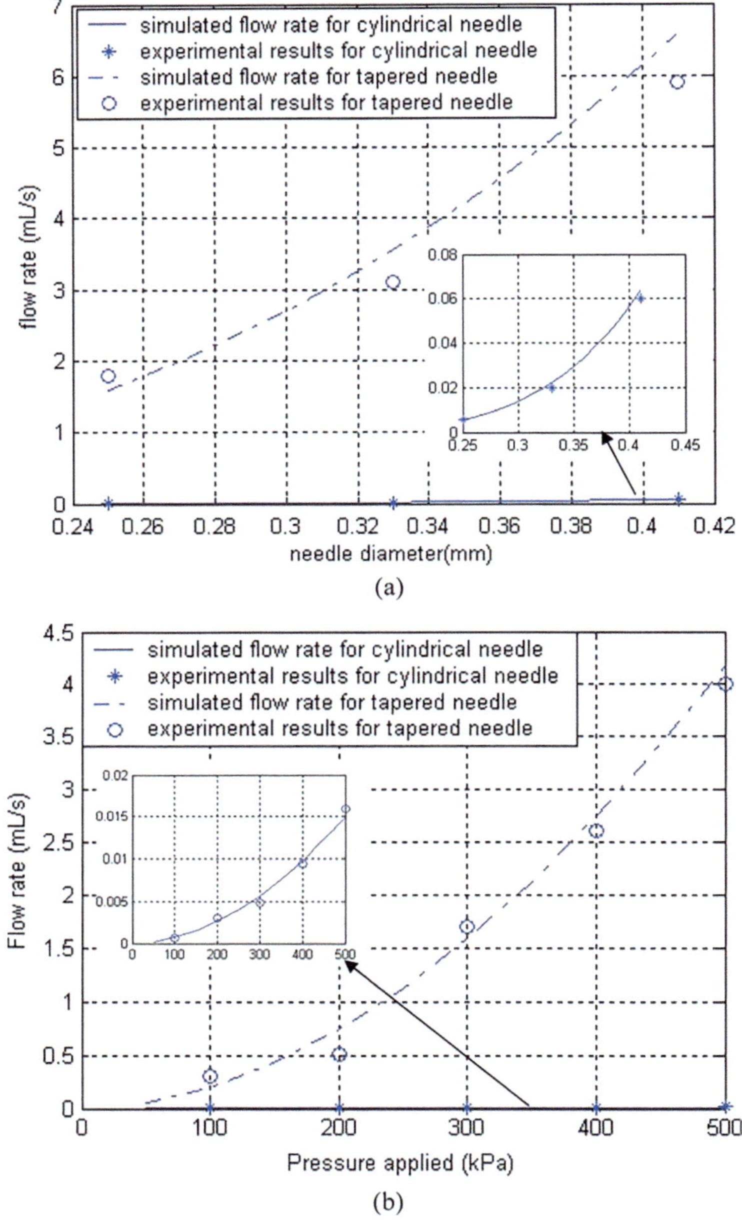

**Fig. 6.11** Comparison of flow rates in cylindrical and tapered needles (**a**) with varying diameters and (**b**) under different air pressures, where embedded figures are enlargements of cylindrical-needle data [7]

comparison of flow rates for two needles with the same exit diameters of 0.25, 0.33, and 0.41 mm under a pressure of 300 kPa; Fig. 6.11b shows the comparison of flow rates for two needles with the same exit diameter of 0.25 mm under pressures of 100, 200, 300, 400, and 500 kPa. Figure 6.11 also shows the flow rates calculated using Eqs. (6.12) and (6.20). The flow rate using the tapered needle is much higher than that using a cylindrical needle under the same printing pressure. For example, at an air pressure of 500 kPa, the flow rate in the tapered needle is almost 200-fold higher. This suggests tapered needles are preferred if high flow rates are required or if the bioink to be printed is viscous and difficult to print using a cylindrical needle. This also suggests that achieving identical flow rates in bioprinting requires a lower air pressure when using a tapered needle vs. a cylindrical needle, which also has the benefit of preserving cell viability in the process.

### 6.3.5    *Extrudability in Bioprinting*

Extrudability in bioprinting is a concept related to the flow rate of the bioink being printed; specifically, it refers to the capability to extrude or print the bioink through a needle to form a continuous and controllable strand. Generally, the printing of bioink from a needle can be characterized as (a) unextrudable, (b) uncontinous (or jetting), (c) continuous yet uncontrollable, or (d) continuous and controllable [11], as shown in Fig. 6.12. In the unextrudable case, the bioink cannot be extruded from the needle; this is typically caused by poor flow behavior (or high viscosity) of the bioink as well as clogging due to large cells and/or their aggregates. Increasing the printing force (e.g., the air pressure in pneumatic-driven bioprinting) can improve extrudability, but may lead to needle breakage or fractured filament morphology. The second case is jetting, where the bioink is extruded into a stream of droplets due to inappropriate

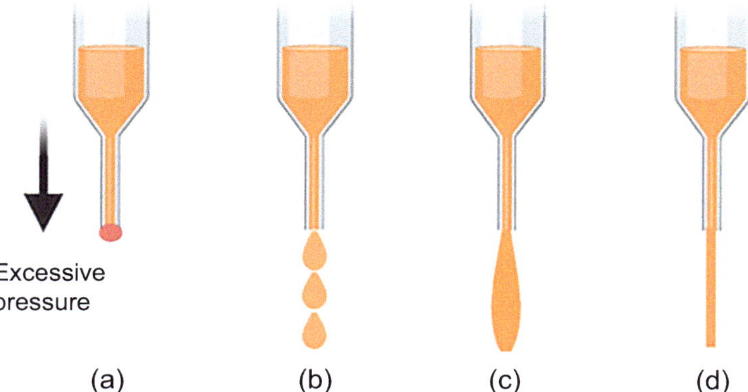


**Fig. 6.12** Extrudability in bioprinting: (**a**) unextrudable, (**b**) uncontinuous (or jetting), (**c**) continuous yet uncontrollable, and (**d**) continuous and controllable

bioink surface tension and flow behavior. In the third case, the bioink can be extruded in the form of filament, but in an uncontrollable manner. In this case, gravity acting on the bioink significantly contributes to the extrusion force but in an uncontrollable way, which is mainly caused by easy-flow behavior (or low viscosity) of the bioink and an internal needle diameter that is too large. In the fourth case, the bioink is extruded both continuously and controllably, and the flow rate of the bioink extruded is well controlled and regulated by the extrusion force.

## 6.4  Bioprinting of Scaffolds

### 6.4.1  Needle Movement in the Horizonal Plane

During the bioprinting of scaffolds, bioink is extruded from the needle, while the needle is controlled to move in the horizonal plane according to the scaffold design. The speed the needle moves in the horizonal plane is important with respect to the cross-section size or profile of the strand formed on the printing stage. Prior to its flowing or spreading, the printed strand has a profile identical to that of the needle. Figure 6.13a, for example, shows a round profile.

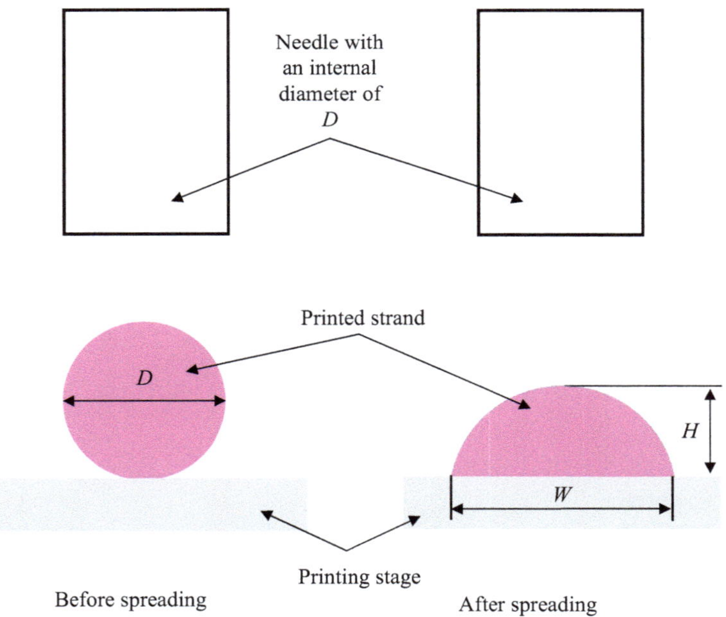


**Fig. 6.13** Profiles (with the same cross-sectional area) of a printed strand before (**a**) and after (**b**) spreading on the printing stage

For a given flow rate, the extruded volume $V_s$ of a bioink within time period $t$ is given by.

$$V_s = Q \cdot t \qquad (6.21)$$

The extruded volume lands on the printing stage as a cylindrical strand with a diameter and length, respectively, denoted by $D$ and $L_d$. As such:

$$V_s = \pi \left(\frac{D}{2}\right)^2 L_d \qquad (6.22)$$

If the horizontal moving speed of the needle is denoted by $v_m$, then.

$$L_d = v_m t \qquad (6.23)$$

From the above equations, we have.

$$v_m = \frac{4Q}{\pi D^2} \qquad (6.24)$$

The above equation demonstrates that the diameter of the printed strand is proportional to the flow rate, depending on the needle speed $v_m$. From this, one can determine the strand diameter:

$$D = \sqrt{\frac{4Q}{\pi v_m}} \qquad (6.25)$$

Equation (6.25) shows that, for a given flow rate, the strand diameter is determined by the needle speed. Note that the use of a particular needle speed will ensure the diameter of the printed strand is equal to the internal diameter of the bioprinting needle if swelling of the material is neglected; such a speed is termed the *stress-free (SF) speed* as there is no stress induced within the strand. If the selected needle speed is faster than the SF speed, the printed strand will stretch to induce a tensile stress within the strand, resulting in a strand diameter smaller than the needle; the converse also holds. Figure 6.14 shows the dependance of printed strands on the needle speed, where the SF speed is about 40 mm/s [4]. If the selected needle speed is too high, the continuity of the strand may not be maintained due to the tensile stress induced within the stand, causing the strand to break. On the other hand, lowering the needle speed below the SF speed induces compressive stress within the strand, leading to irregular stand orientation and increased strand diameter. If the needle speed becomes too low, printing a straight strand becomes difficult due to the induced compression.

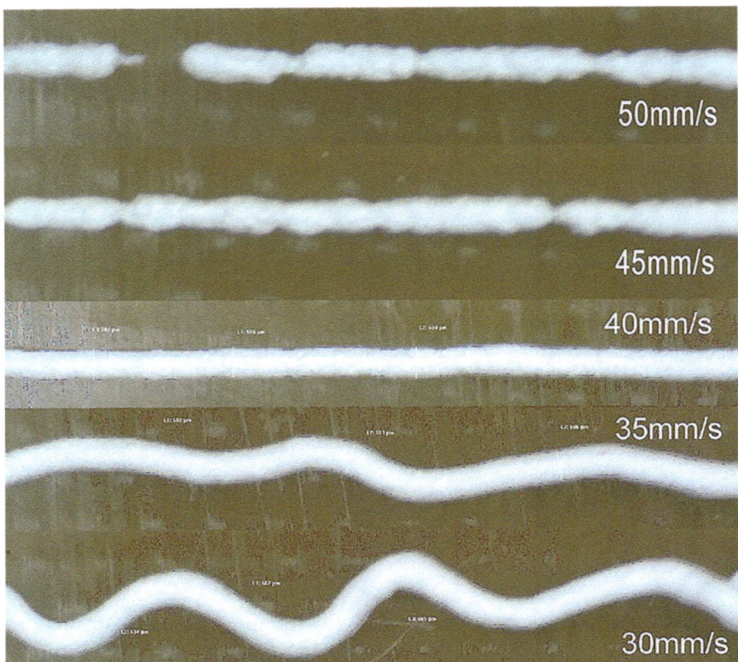


**Fig. 6.14** Effect of needle speed on the printed stands, where the SF speed is around 40 mm/s [4]

In addition to strand formation, needle speed also influences the molecular structure of printed strands (mainly due to the induced stress within the strand) and thus the biological behavior of scaffolds. Creating molecular structures able to direct anisotropic cell growth in scaffold fabrication is crucial to mimic the structure and function of native tissues in some tissue engineering applications. In nerve tissue engineering, for example, scaffolds with linearly arrayed extracellular components and Schwann cells are preferred to support the orientation and growth of neuronal axons along the nerve. Studies on bioprinting bioinks containing fibrin and Schwann cells for nerve scaffolds have illustrated that fibrin fibers with aligned morphology can be introduced by controlling the bioprinting flow rate and needle speed [12, 13]. This is because the molecular structure of fibrin enables fibrin fiber networks to undergo reorganization with induced stresses. By increasing the needle speed to exceed the SF speed, the polymeric structure is aligned due to the induced tensile stress within the printed strand in the direction of needle movement. These physical guidance cues of aligned fibrin fibers result in the linear alignment of Schwann cells encapsulated inside the strands, which mimics the morphology of nerve tissue cells and thus contributes to directional neurite outgrowth [12, 13].

## *6.4.2  Needle Movement in the Vertical Direction*

As a scaffold is a 3D structure stacked from multiple layers, the control of needle movement in the vertical direction is also important to the stability and fidelity of the final printed scaffold. The first layer of bioprinting provides an anchor to hold subsequent layers and the whole scaffold to the printing stage. The distance between the needle tip and the printing stage for the first layer (denoted by $d$ in Fig. 6.13) is particularly important. If the space between the outlet of the needle tip and stage is too large, dripping droplets or segments might form at the needle tip due to the action of the surface tension of the bioink. This interrupts the continuity of strand formation in bioprinting [14], leading to an inconsistent diameter in the printed strand. On the other hand, if the needle tip is too close to the printing surface, the printed strand will be scratched by the moving needle, and the strand formed can have a large or even uncontrollable width (denoted by $W$ in Fig. 6.13b). In general, a distance that is slightly larger than the inner diameter of the needle is recommended when printing the first layer.

Ideally, printed strands are cylindrical in shape with a diameter equaling the inner diameter of the needle. However, this is not actually the case due to the flow or spreading of bioink caused by the effect of gravity, especially when soft materials such as hydrogels are used, as well as swelling of the bioink. Bioink spreading will lead to the strand dimension reduced in the vertical direction, while increased in horizonal direction, as shown in Fig. 6.13b). To mitigate the dimensional reduction or deflection of printed strands, strategies for toolpath control can be developed to partially compensate for the deflection of strands in the vertical direction. Generally, the movement of the needle in the vertical direction is set as a certain percentage of the needle diameter by considering the spreading of bioink (referred to the predictive compensation). However, due to the substantial weight of printed strands as more layers are stacked, the distance between the needle and the printing layer is prone to increase over time. To address this issue, strategies with predictive compensation regarding the deflection of scaffolds in the vertical direction are promising as they can automatically adjust the distance between the needle tip and the printing layer when printing different scaffold layers.

To reduce the flow or spreading of bioink when it is printed on a stage, one effective way is to print bioink into a medium to create scaffolds. Different from printing in air, printing into a medium means buoyancy acts on the printed bioink and counteracts the influence of gravity; as a result, bioink flow or spreading on the print stage or in the medium can be reduced, as shown in Fig. 6.15. The use of a crosslinking agent solution (in Sect. 6.4.3) and a supporting medium in embedded bioprinting (in Sect. 6.7.1) are examples of printing bioink in a medium. By doing so, control of movement in the $Z$ direction might become much simpler as a fixed distance for each layer can be used without the need for preset compensation as discussed above.

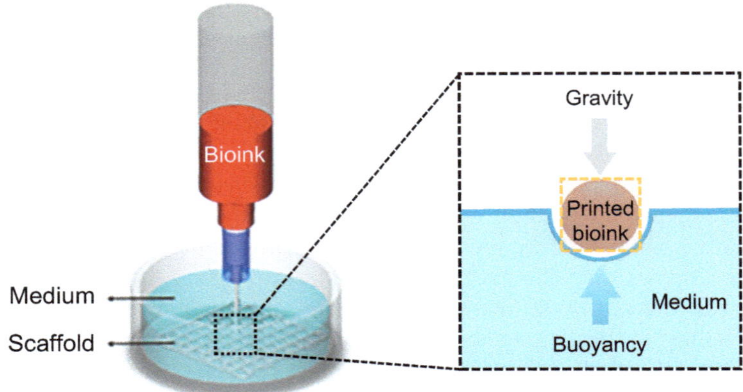


**Fig. 6.15** Printing bioink into a medium, with buoyancy counteracting the influence of gravity

## 6.4.3 Crosslinking in Bioprinting

To create hydrogel-based 3D scaffolds with integrated structures, the hydrogel solution must be solidified or crosslinked to enhance the mechanical strength and stability. Crosslinking can be initiated by means of physical stimuli or chemically induced via a crosslinking agent or enzymatic reaction [15]. Among such methods, the most commonly used for bioprinting are ionic, thermal, and photo crosslinking, as discussed in Sect. 3.2.2. Depending on hydrogel properties, various crosslinking methods with associated configurations have been developed for bioprinting, as shown in Fig. 6.16.

By manipulating the temperature of thermo-sensitive bioinks, such as those based on agarose, gelatin, or collagen, before and after printing, thermal crosslinking of the bioink can take place in the bioprinting process. Figure 6.16a shows a configuration where the temperature of the bioink in the syringe is controlled above the melting temperature of the bioink to retain its solution form; the temperature of the printing stage is then set to below the gelation temperature so that the bioink, once printed, gels on the stage. As such, the crosslinking process requires a temperature-regulation controller installed in the bioprinting system.

Ionic crosslinking is reversible and has been extensively applied for hydrogels such as alginate and chitosan. Gelation occurs upon the formation of ionic bonds after the polymer molecules encounter the crosslinking agent. Chemical crosslinking is similar to ionic crosslinking, where hydrogel solutions and crosslinking agents must come into contact. The formation of hydrogels is triggered by the crosslinker, which connects hydrogel molecules via covalent chemical bonds. One approach developed to introduce crosslinking agents is to atomize and then spray them onto the extruded hydrogel solution (Fig. 6.16b). Challenges related to this method include the control of atomized agents to allow homogeneous distribution on extruded solutions to form strands with uniform diameter, as well as relatively

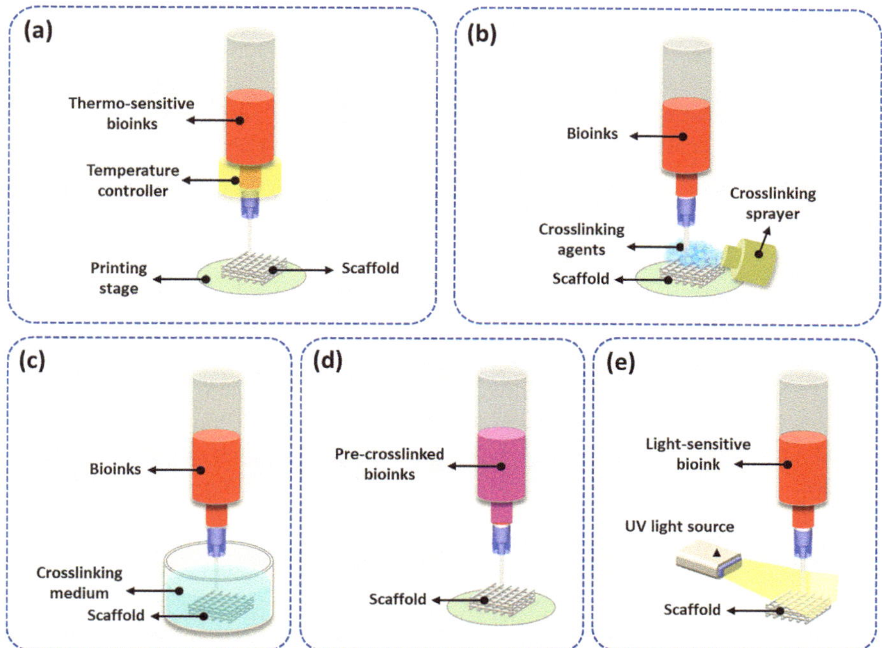


**Fig. 6.16** Methods and configurations for hydrogel scaffold crosslinking: (**a**) crosslinking under temperature control; (**b**) crosslinking under spray; (**c**) crosslinking in medium bath; (**d**) pre-crosslinking; and (**e**) crosslinking under a UV light

slow gelation and incomplete crosslinking due to the limited effectiveness of the atomized crosslinking agent. Thus, maintaining structural fidelity and stability becomes difficult, especially when hydrogel precursors with poor mechanical properties such as low viscosity are utilized. To address these issues, hydrogel solutions can be deposited into a bath containing crosslinking medium (Fig. 6.16c). As sufficient crosslinking agents are homogeneously provided in the bath, the surface of the deposited solution that contacts the crosslinkers is rapidly solidified, limiting the spread of the hydrogel solution and thus supporting the fidelity of printed strands. This method is also known as 3D bioplotting and requires the careful adjustment of crosslinking agents because the buoyancy of the crosslinking medium may lead to the failure of scaffold stacking if an inappropriate crosslinking solution is utilized. An excessive gelation rate introduced by a high concentration of crosslinking agent would result in rapid stiffening of the strand surface, which may reduce the connection between adjacent layers and lead to poor scaffold stability. On the other hand, slow gelation speeds caused by a low concentration of crosslinker result in poor fidelity of the printed strands due to solution spreading as well as poor mechanical properties and even failure to support the printed structure [16]. Pre-crosslinking, by adding and mixing low concentrations of crosslinking agent into the hydrogel solution, can also be applied for hydrogel-based bioprinting (Fig. 6.16d). The pre-crosslinking method introduces hydrogel particles in the hydrogel solution, which increases its viscosity and therefore the deposition quality. The structural stability of printed scaffolds can then be easily achieved by exposing the printed scaffold to a high concentration of crosslinker solution. The mechanical properties of pre-crosslinked scaffolds are good, but the printing pressure required during bioprinting increases relative to the density of the pre-crosslinked hydrogel. In addition, pre-crosslinking introduces an uneven distribution of hydrogel particles in the hydrogel solution, leading to discontinuities and nonuniformities during extrusion.

A UV light beam can be used in bioprinting to initiate hydrogel photopolymerization (Fig. 6.16e). A hydrogel can be photopolymerized in the presence of photoinitiators under lights. When the light source is used, the interaction of the light source and light-sensitive compounds (photoinitiators) initiates polymerization. For example, gelatin is an inexpensive, denatured collagen that retains an abundance of integrin-binding motifs and matrix metalloproteinase-sensitive groups that promote adhesion of cells. By adding methacrylate and methacrylamide groups to the amine-containing side groups, gelatin becomes gelatin methacryloyl (GelMA), a photopolymerizable material. GelMA maintains the thermosensitive properties of gelatin and not only can be gelled under temperature control but can also be permanently polymerized upon exposure to UV light. As such, extrusion-based bioprinting in combination with a UV light source can be used to produce a GelMA-based scaffold.

### *6.4.4  Design of Bioprinting Parameters*

Some parameters are important for the bioprinting process, such as the pressure in pneumatic-based bioprinting and the needle movement speed in the horizonal plane; these parameters need to be designed or determined with appropriate values for use in the bioprinting process. Experience-based or trial-and-error methods can always be used for this purpose. Instead, this section presents methods, based on the models presented in the proceeding sections, to rigorously design or determine the bioprinting parameters through an example, where the bioprinting parameters of the pressure and needle speed are determined from the scaffold design as well as the flow behavior of bioink prepared for printing.

**Example 6.2** Consider the scaffold design shown in the following figure, where the strand diameter is 200 μm and the strand spacing is 600 μm. Suppose that the scaffold is to be printed by a pneumatic-driven bioprinting process from a bioink with flow behavior described by $\tau = 10\dot{\gamma}^{0.4}$. Determine

(continued)

**Example 6.2** (continued)
the pressure and needle speed for use in the bioprinting process for the case in
which no stretch or compression occurs in the printed strands.

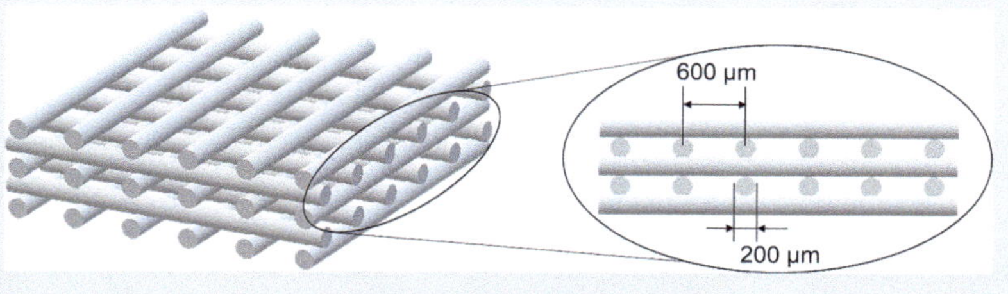


**Solution**
*Select the needle for bioprinting.* Based on the strand diameter, select the needle with
an inner diameter equal to the strand diameter if ignoring bioink swelling. For the
given scaffold design in this example, select a 27-gauge needle (color: clear) with a
length of 12 mm (from Fig. 6.10).

*Method 1: Select the pressure and then determine the needle speed.* Select one
pressure from the range available in the pneumatic-based bioprinting system and
suppose it is $\Delta P = 60$ kPa. By ignoring the influence of bioink surface tension, the
flow rate of the bioink printed is, from Eq. (6.12):

$$Q = \frac{n\pi R^{(3n+1)/n}}{(3n+1)(2KL)^{1/n}}(\Delta P)^{1/n}$$

$$= \frac{0.4\pi\left(100\times 10^{-6}\right)^{(3\times 0.4+1)/0.4}}{(3\times 0.4+1)\left(2\times 10\times 12\times 10^{-3}\right)^{1/0.4}}\left(60\times 10^{3}\right)^{1/0.4} = 1.79\times 10^{-9}\ \text{L/s}$$

The needle speed can be determined from the deposition speed, which is calculated from Eq. (6.24), i.e.,

$$v_m = \frac{4Q}{\pi D^2} = \frac{4\times 1.79\times 10^{-9}}{\pi\times\left(200\times 10^{-6}\right)^2} = 56.97\ \text{mm/s}$$

If the above needle speed is too large or small for bioprinting in terms of the range
of needle speed available on the printer, decrease or increase the selected pressure
until the calculated needle speed is appropriate for the system employed.

*Method 2: Select the needle speed and determine the applied pressure.* Select the
needle speed from the range available in the pneumatic-based bioprinting system;
suppose it is 40 mm/s. From Eq. (6.24), the flow rate of bioink printed can be
calculated:

$$Q = \frac{\pi D^2}{4} V_m = \frac{\pi \times \left(200 \times 10^{-6}\right)^2}{4} \times 40 \times 10^{-3} = 1.26 \times 10^{-9} \ \text{L/s}$$

From Eq. (6.12), the applied pressure can then be determined:

$$\Delta P = \frac{2KL(3n+1)^n}{(n\pi)^n R^{(3n+1)}} Q^n = \frac{2 \times 10 \times \left(12 \times 10^{-3}\right) \times (3 \times 0.4 + 1)^{0.4}}{(0.4 \times \pi)^{0.4} \times \left(100 \times 10^{-6}\right)^{(3 \times 0.4 + 1)}}$$

$$\times \left(1.26 \times 10^{-9}\right)^{0.4} = 5.22 \times 10^4 \ Pa$$

If the above applied pressure is too large or small with respect to the range of pressure available on the printer, decrease or increase the needle speed selected above until the calculated pressure is appropriate for bioprinting.

## 6.5   Profile and Structure of Printed Scaffolds

### 6.5.1   Profile and Structure

During the bioprinting process, bioink is deposited or printed on the printing stage as discussed in the sections above. However, the bioink, once printed on the stage, is still in a solution or semi-solution form. Due to the effect of gravity, the bioink can flow or spread, thus mixing or fusing at the intersection of strands or deforming where upper strands span the gap between lower strands or hang over at their ends. As a result, the strand profile and scaffold structure become different from that designed, as shown in Fig. 6.17, where the top view (a) illustrates the variation in strand cross-section with location and the cross-sectional view (b) illustrates the deflection of strand and pore size in the vertical direction as well as the reduced height of the whole construct.

### 6.5.2   Techniques to Characterize Profile and Structure

Various techniques have been adapted or developed to characterize the strand profile and scaffold structure in terms of scaffold pore size, pore volume/ porosity, and pore interconnectivity of fabricated scaffolds. Generally, these techniques can be classified as non-imaging or imaging approaches.

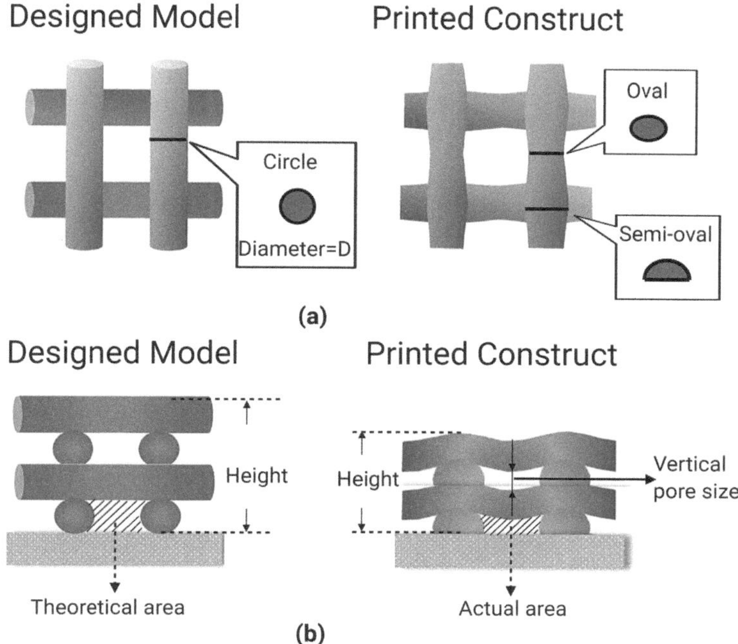


**Fig. 6.17** Difference in strand profile between designed and printed scaffold structures: (**a**) top view and (**b**) cross-sectional view

### 6.5.2.1   Non-imaging Approaches

A profile-measurement system can measure the 2D profile of an object (or printed bioink) by mean of a laser, such as the one shown in Fig. 6.18. This system integrates a laser emitter and laser receiver, which are controlled to move over the object being scanned or measured. During the measurements, the system scans the object surface by emitting a laser beam and receiving its reflection, from which it then infers the object profile. The profile-measurement system can be used to measure the profile of printed bioink as well as the contact angle of the bioink on the printing stage.

Gravimetry is one approach to measure the porosity of a scaffold that was developed based on the definition of scaffold porosity, and expressed as a percentage,

$$\%Porisity = 1 - \frac{\rho_{total}}{\rho_{material}} \tag{6.26}$$

where $\rho_{total}$ is the apparent density of the scaffold measured by dividing the weight of the scaffold by the total volume, and $\rho_{material}$ is the density of the material from which the scaffold was created.

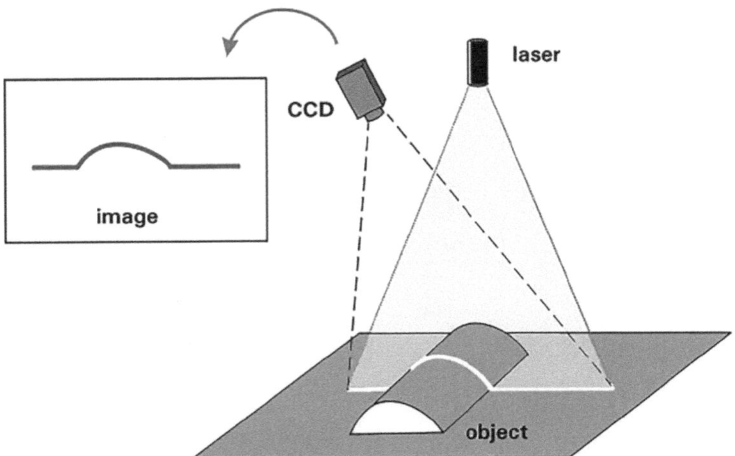


**Fig. 6.18** Characterizing the profile of printed bioink by a laser scanner

Mercury porosimetry is a technique to measure the pore size of a scaffold based on the pressure required to force mercury into the pores against the resistance of liquid surface tension. The measurements from this technique give indicative, rather than actual, values of the pore size and allow for the characterization of pore size in a wide range from a few nanometers to several hundreds of micrometers. A limitation of this method is the poor resolution when large pore sizes (over 500 µm) are measured because low mercury intrusion pressures are necessary. This method is also limited in terms of applications to scaffolds with irregular pore geometries.

### 6.5.2.2   Imaging Approaches

Imaging techniques, such as micro-computed tomography (µCT), have been utilized to characterize the porosity and pore size of printed scaffolds. µCT is a computer-processing and X-ray-based imaging technique to produce cross-sectional images of scanned scaffolds at a small scale [17, 18]. During scanning, the scaffold is placed in the µCT device and subjected to X-ray scanning; the isotropic slice data obtained from the scan are then rebuilt into 2D images. With the series of 2D images obtained, 3D visual structures including compiled morphological detail of the scaffold are reconstructed (Fig. 6.19). Utilization of µCT techniques directly provides structural information such as pore location and pore size; based on image processing and analysis, the scaffold porosity, pore interconnectivity, and pore distributions can be characterized. µCT techniques allow for more flexibility in terms of measurements and also provide more information compared to gravimetry and mercury porosimetry techniques. The morphology of the pore structures of printed scaffolds made from hydrogels has been estimated using µCT scans [17–20]. The

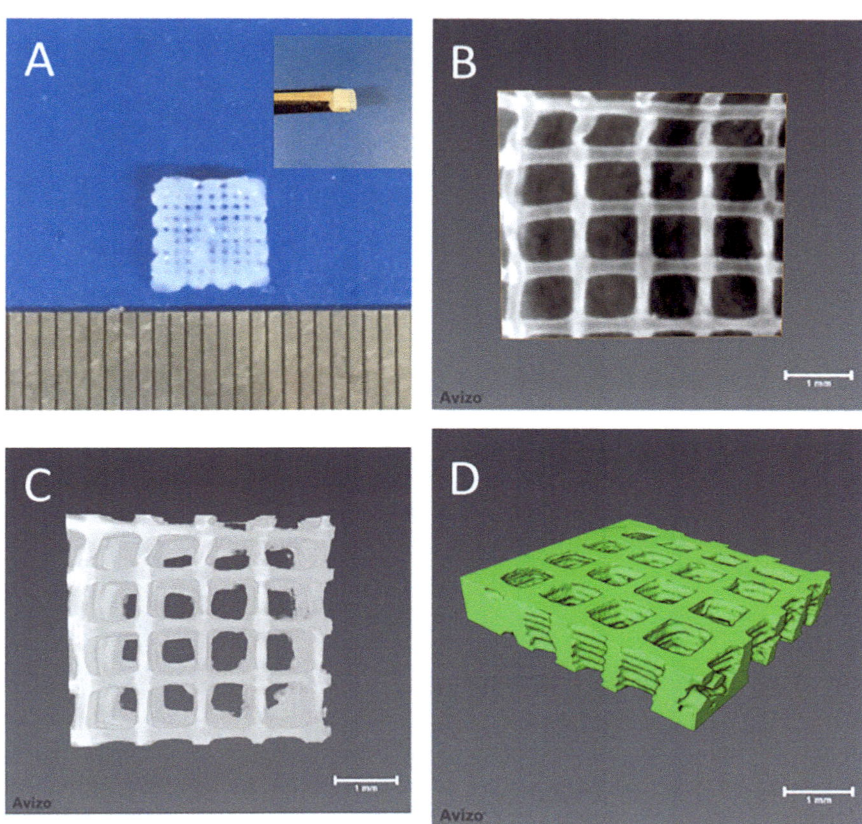

**Fig. 6.19** Printed scaffolds and morphology evaluation: (**a**) printed scaffold with a 0–90° inner structure; (**b**) phase-retrieval slice of printed scaffold after CT scan; (**c**) top and (**d**) side views of the 3D scaffold after reconstruction [20]

reconstructed images show that, for hydrogel-based printing, both the horizontal and vertical pore sizes can be significantly different than design models due to the spreading of materials and the fusion of adjacent layers. In addition, the precision of porosity and pore interconnectivity relies heavily on the properties of the materials used, bioprinting process control, as well as printing method selected. Utilization of μCT requires post-treatment of hydrogel scaffolds (i.e., lyophilization), which might change the structure and therefore degrade the accuracy of pore size and interconnectivity evaluations [20].

Scanning electron microscopy (SEM) is another commonly used imaging technique to characterize scaffold pores and porosity. SEM captures sample images by scanning the surface with a beam of electrons. By interacting with the atoms on the sample surface, the electrons produce signals associated with the surface topography and composition, which are then detected to produce an image. From the SEM

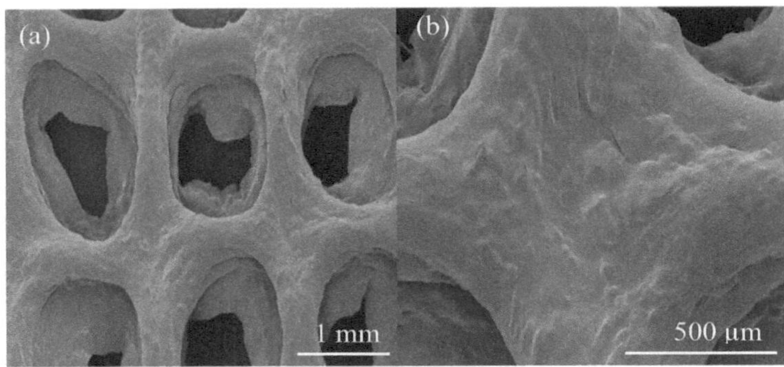


**Fig. 6.20** SEM images of a printed chitosan scaffold after air-drying [21] with different magnifications

images, the pore size and pore interconnectivity can be analyzed and characterized. Figure 6.20 shows an example of the SEM images of a scaffold (after air-drying) [21], which are 2D and display the scaffold internal structure, pore size, and pore interconnectivity.

## 6.6 Cell Damage in Bioprinting

Bioprinting allows for the incorporation of living cells within scaffolds, but the printing process may also damage the cells incorporated. Living cells are dynamic structures and their functioning (e.g., growth and proliferation) can be affected by mechanical forces they experience. Mechanical forces are induced during the bioprinting process and can cause the deformation and breach of cell membranes of incorporated cells. Although cells have elastic abilities to resist a certain level of force, cell membranes may lose their integrity if the applied force exceeds a certain threshold; as a result, cells may be damaged and lose their functionality [22–26]. Therefore, understanding the process-induced forces on cells during the bioprinting process becomes important for preserving cell viability in scaffold bioprinting.

### 6.6.1 Process-Induced Mechanical Forces

Mechanical forces are induced in the bioprinting process, including compressive force, shear stress, and extensional stress. The compressive force on cells is generated due to hydrostatic pressure when cells are suspended in solution (Fig. 6.21a). During bioprinting, the compressed air creates forces on the cell suspension loaded in the syringe, with the corresponding hydrostatic pressure approximately equaling

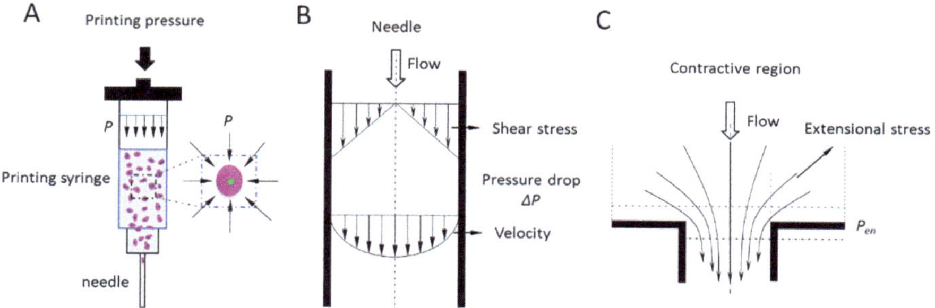


**Fig. 6.21** Mechanical forces to which cells are subjected during bioprinting: (**a**) hydrostatic pressure; (**b**) shear stress (for simplicity, $\tau_0 = 0$); and (**c**) extensional stress

the bioprinting pressure (if the pressure drop in the syringe can be ignored). Hydrostatic pressure is also present in the needle, the magnitude of which is dependent on the location of cells inside the needle, as given by

$$P_n(l) = \left(1 - \frac{l}{L_n}\right)\Delta P \tag{6.27}$$

where $\Delta P$ is the pressure drop along the needle, $L_n$ is the needle length, and $l$ is the distance from the needle entrance to the location of cells inside the needle ($0 < l < L_n$).

Shear stress is a mechanical force that introduces cell damage during bioprinting. Because the diameter of the syringe is much greater than the needle tip diameter, the bioink flow inside the syringe can be neglected and therefore the bioprinting process-induced shear stress is predominantly distributed inside the narrow needle tip as the cell suspension is forced to flow through (Fig. 6.21b). Equation (6.4), which is reproduced here,

$$\tau = \left(\frac{r}{2}\right)\left(\frac{\Delta P}{L}\right)$$

shows that shear stress in the needle tip is dependent on the pressure drop and the length of the needle, and is linearly distributed along the radial direction inside the needle considering a fully developed flow and no slip at the needle wall. For a given pressure and needle length, the shear stress reaches a maximum value at the needle wall and decreases to zero at the center of the needle.

Another mechanical force to which cells are subjected is extensional stress, which is a tensile stress generated due to the extensional flow field. Extensional flow is induced at the region of abrupt contraction of the needle, where the solution velocity difference before and after the contractive region is large (Fig. 6.21c). To express the extensional stress in the contractive region, one approach has been developed based on the pressure drop in the coni-cylindrical entrance due to shear flow and extensional flow and assuming these terms are additive [27]. Based on this assumption, the fluid would adopt a contractive profile so as to minimize the total pressure drop, with the fluid having an extensional viscosity independent of the extensional rate while the shear viscosity is a power-law function of the shear rate. Thus, the expression for the extensional stress is.

$$\tau_e = \frac{3}{8}(n+1)P_{en} \tag{6.28}$$

where $\tau_e$ is the extensional stress, $P_{en}$ is the pressure drop at the contractive region, and $n$ is the power-law index for shear flow. This expression is only applicable in situations in which the entrance angle of the entrance region is sufficiently large so as to not interfere with the flow pattern. This can be only guaranteed for large angles that approach an abrupt or flat entry with an entrance angle of 90°. Also, the resulting extensional stress is an average value that cannot capture features of the flow from different streamlines in the region.

Another method expresses the entrance flow based on a power-law model, which enables calculation of the extensional stress and extensional rate [28]. It incorporates more complex rheological behavior into the analysis by allowing the extension viscosity to depend on the extensional rate, as follows:

$$\tau_e = s\dot{\gamma}_e^m \tag{6.29}$$

where $s$ and $m$ are power-law indexes similar to the model used to describe the shear stress, and $\dot{\gamma}_e$ is the extensional rate. This method can predict the extensional stress at each flow streamline according to the extensional rate; however, unlike a Newtonian fluid, for which the extensional viscosity is constant and three times the steady shear value, it is experimentally difficult to produce the pure normal flow fields that are essential for identifying the flow indexes $s$ and $m$. In addition, the flow pattern at the contractive region of the needle is normally complex, leading to difficulties in assessing and quantifying the extensional stress and its distribution.

> **Example 6.3**  A cell/hydrogel bioink is printed at room temperature through a 100-μm diameter, 12-mm long cylindrical needle on an extrusion-based bioprinting system. The printing pressure is 100 kPa and the flow inside the needle is fully developed.
>
> 1. The diameter of the printing syringe is much greater than the diameter of needle. Determine the hydrostatic pressures to which the cells are subjected for different positions in the printing system as shown in the figure below (do not consider the pressure drop due to the contractive region).

(continued)

**Example 6.3**  (continued)

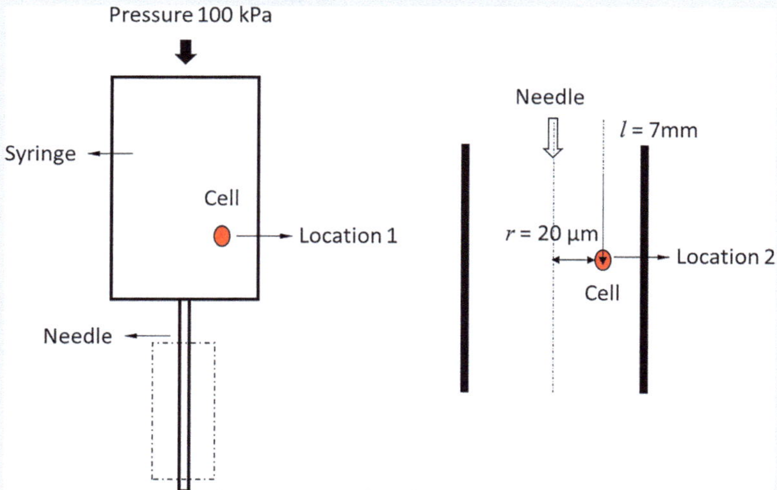


2. Determine the shear stress to which the cells are subjected inside the
   needle (location 2) and calculate the time for a cell to travel through the
   needle if it follows the vertical streamline from the inlet to the outlet of the
   needle (suppose the power-law index $K$ is 23.5 and $n$ is 0.45).

## Solution

1. For the cell in the syringe (location 1), the hydrostatic pressure approximately
   equals the bioprinting pressure considering that the pressure drop in the syringe
   can be ignored because the syringe diameter is much greater than the needle
   diameter. The given air pressure is 100 kPa, so the hydrostatic pressure on the cell
   is 100 kPa.

   For the cell inside the needle (location 2), the pressure to which it is subjected
   can be calculated according to Eq. (6.14):

$$P_n(l) = P - \frac{l}{L_n} \Delta P_n$$

where $P$ is the air pressure (100 kPa) and $\Delta P_n$ is the pressure drop in the needle,
which equals the given pressure if the pressure drop at the contractive region is
neglected. As $l = 7$ mm and $L_n = 12$ mm, the pressure to which the cell is exposed
is 41.67 kPa.

2. The shear stress experienced by the cell located inside the needle can be expressed based on Eq. (6.2), where the shear stress is related to the radial position:

$$\tau_{rz} = \left(\frac{r}{2}\right)\left(\frac{\Delta P}{L}\right)$$

As $r = 20$ µm, the shear stress to which the cell is subjected is 83.33 Pa.

If the cell follows the vertical streamline from the inlet to the outlet of the needle, based on Eq. (6.5),

$$V_z = \left(\frac{n}{n+1}\right)\left(\frac{\Delta PR}{2KL_n}\right)^{1/n} R\left[1 - \left(\frac{r}{R}\right)^{\frac{n+1}{n}}\right]$$

the velocity is 1.877 mm/s. As the length of the needle is 12 mm, the travel time of the cell is 6.39 s.

### 6.6.2 Cell Damage Due to Mechanical Forces

Cells are the basic structural and functional units of living organisms; they consist of cytoplasm enclosed within a membrane and generally have a spherical morphology when suspended in solution. Under mechanical forces, cells are passively deformed to reach a new morphological state. Once the forces exceed certain levels, cells will be injured and damaged to the point that they cannot maintain their phenotype and cell membrane integrity, leading to cell damage or a reduction in cell viability. Research has demonstrated that cell viability during bioprinting depends on the bioprinting mechanisms, process-induced mechanical forces, and cell types [22–26].

During bioprinting, cells inside the syringe and needle are subjected to hydrostatic pressure, which can change their morphology by compressing the cell membrane to a new state (Fig. 6.22, left panel). The effects of bioprinting pressures on cell damage have been experimentally investigated on various cell types, including Schwann cells, fibroblasts, and chondrocytes [22]. The results demonstrate disruptive effects of compressive pressure on cell cytoskeletal organization and membranes only if the pressures reach high levels (e.g., 5 MPa) and are maintained for a long period (e.g., 2 h). As bioprinting normally requires much lower printing pressures and shorter times, it is commonly accepted that the effects of hydrostatic pressure are minor; thus, the cell damage that occurs in the bioprinting process can be primarily attributed to other mechanical forces, including shear and extensional stresses [26].

Cells are oriented and then deformed into another state when exposed to shear flow (Fig. 6.22, center panel). The shear stress twists and stretches the cell membrane

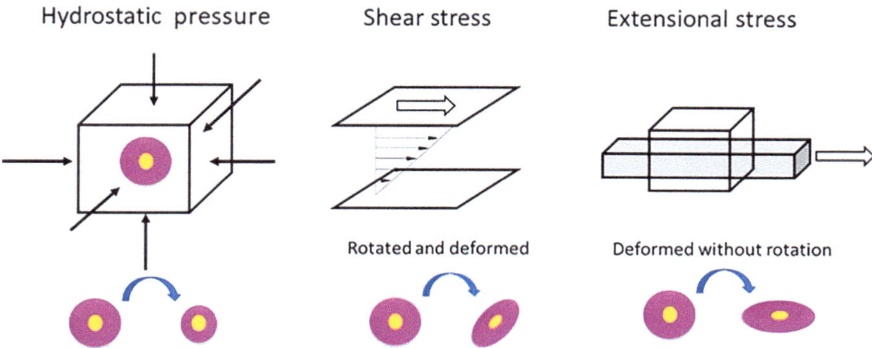



**Fig. 6.22** Deformation of cells due to bioprinting process-induced mechanical forces, including hydrostatic pressure, shear stress, and extensional stress

and, if some physiological threshold is reached, cell damage occurs. An analysis of the bioprinting process-induced shear stress distribution inside the needle shows larger cell deformations and more damage to cells in flow streamlines close to the needle wall as opposed to those near the needle center [22].

Cells experience high extensional stress when they pass through the contractive region of the needle. Compared to shear stress, the extensional stress directly stretches cells, deforming them in the stretching direction without rotation (Fig. 6.22, right panel). Cells are easily damaged by this mechanism. Extensional stress has been identified as a major cause of acute cell damage and, therefore, reducing the magnitude of extensional stress or protecting cells from extensional stress during bioprinting is crucial for preserving cell viability [26].

### *6.6.3   Cell Damage and Its Characterization in Bioprinting*

Cell damage during bioprinting varies depending not only on the process-induced forces that cells experience (as discussed in the preceding section) but also on the cell types due to the inherent, yet distinct, ability of their structures and properties to endure the process-induced forces. Methods to characterize cell damage during bioprinting can generally be classified as experimental or modeling/simulation.

The bioink properties (e.g., flow behavior) and printing process parameters (e.g., printing pressure, needle length, and needle diameter for cylindrical needles) affect the process-induced forces, and thus cell damage, in the bioprinting process. Experimental characterization is a straightforward method to examine the cell damage in bioprinting by measuring the cell viability (with the assays discussed in the following section) and correlating it to bioink properties and printing process parameters. Larger numbers of damaged cells are observed when higher printing pressures are applied due to larger induced shear and extensional stresses inside the printing needle [26]. Using low air pressures can reduce the degree of cell damage but might be impractical for printing viscous bioinks. Larger pressures are required to achieve a given flow rate for printing more viscous solutions but result in greater shear and extensional stresses on cells and lead to increased cell damage. For a given printing flow rate, less cell damage occurs when a larger-diameter needle is applied; this is attributed to the lower pressure required to maintain the flow rate. Although a larger-diameter needle can be used to reduce the pressure if a certain flow rate is required, the printing resolution and accuracy can be reduced, thus precluding some precision applications. All of these experimental observations suggest that the printing pressure and needle size for bioprinting can be determined and optimized to minimize cell damage.

In parallel with experimental methods, modeling and numerical simulations have also been developed to characterize bioprinting process-induced cell damage, resulting in various models that aim to correlate process parameters to the degree of cell damage. A phenomenological model has been used to describe the degree of cell damage as a function of the needle employed and dispensing pressure [29]. The model fits the degree of cell damage with experimentally identified model coefficients for given cell types and printing conditions. However, evaluating cell damage and viability corresponding to needle diameters and air pressures fails to capture the direct mechanism of mechanical stresses on cells. Another modeling method to quantify cell viability in the bioprinting process is based on three systematic steps: (1) developing empirical models for cell damage or cell damage laws, which relate the percent cell damage to the mechanical force of hydrostatic pressure or shear/extensive stress that cells experience (similar to Newton's second law that relates an object's motion to the mechanical force applied), (2) evaluating the mechanical forces that cells experience in every cell path in which they flow in the bioprinting process, and (3) determining the percent cell damage in all cell paths based on the results from the above steps, and then integrating them to obtain the overall percent cell damage in the bioprinting process and thereby the cell viability. To develop the cell damage laws required in step (1) above, Fig. 6.23, as an example, shows the

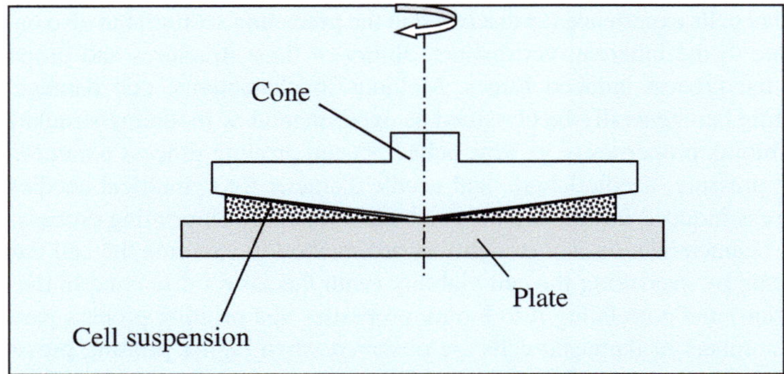


**Fig. 6.23** Schematic of an experimental setting to examine cell responses to shear stress

schematic of an experimental setting in which a plate-and-cone rheometer is used to examine cell responses to shear stress, thus establishing cell damage laws under shear stress. Specifically, a cell suspension is sheared under torsion within the rheometer gap by the rotating cone, and the cell damage then measured in terms of the magnitude of shear stress and shear time or duration. Empirical cell damage models or cell damage laws describing the relationship between cell damage and shear stress and shear time can be established and thus used to characterize bioprinting process-induced cell damage.

In addition to the printing pressure and needle size discussed above, the degree of cell damage is also associated with such factors as the needle shape (i.e., cylindrical or tapered) [7] and temperature [30]. Experimental results demonstrate that tapered needles require much less pressure than cylindrical needles if a constant flow rate is required, resulting in lower induced stresses and a reduced degree of cell damage. However, tapered needles face issues related to excessive and uncontrolled flow rates for low-viscosity solutions, even at very low printing pressures. In precision applications, for example, where viscous materials are used to print structures with high-resolution strands, a tapered needle with a small inner diameter can be selected to ensure printing precision while preserving cell viability. An example with Schwann cells indicates that bioprinting temperature also influences cell damage [30]; specifically, Schwann cells at similar levels of stress are more easily damaged at higher temperatures. Cell damage initially increases slowly and then accelerates at temperatures of 15 °C or more at certain shear stresses. As such, the reliability of cell damage characterization can be further improved by considering the influence of temperature in addition to other bioprinting process parameters. This helps to understand cell damage mechanisms and optimize the viability of cells in scaffold bioprinting.

With more evidence indicating that extensional stress plays a key role in cell damage during bioprinting [26], its effects should be considered in cell damage characterization. Human umbilical vein endothelial cells (HUVEC), human adipose stem cells, rat mesenchymal stem cells, and mouse neural progenitor cells have been used in the development of cell models to illustrate the dominant influence of extensional stress on cell damage. As the maximum shear stress that occurs at the wall of the printing needle can be calculated, that shear stress can be directly applied and controlled using a rheometer. Viability tests for linearly sheared cells show that more than 90% of cells are alive after shearing, which is much higher than the viability of printed cells. The additional cell damage is believed to be induced by extensional stress. Due to the complexity of extensional flow at the contractive region, the relationship between cell damage and extensional stress during bioprinting has not been well documented. While the effects of shear stress on cell damage can be investigated using a plate-and-cone rheometer (Fig. 6.23), assessing and quantifying cell damage caused by extensional stress during bioprinting is challenging because producing a pure normal flow field with a cell suspension is difficult, leading to issues with respect to experimentally characterizing the extensional stress-induced cell damage. One method is to indirectly establish an extensional stress-induced cell damage model by considering both shear and extensional stresses in bioprinting [26]. The bioprinting process-induced cell damage can be considered an aggregation of both shear and extensional stress-induced cell damage. As the percentage of cell damage attributable to shear stress can be determined and the total cell damage can be easily measured after printing, cell damage due to extensional stress can therefore be calculated as the difference. Thus, an extensional stress-based cell damage law can be established once the extensional stress profile is determined.

### 6.6.4   Cell Viability Assays

Cell viability assays are used to measure and determine the proportion of healthy cells within samples, and various assays have been adopted to examine the cell viability within printed constructs. To distinguish the damaged cells from normal cells, dyes such as azo dye trypan blue or fluorescent dyes calcein-AM and propidium iodide have been widely used due to their ability to selectively stain live and dead or damaged cells. Trypan blue is used to selectively color damaged or dead cells blue. Cells with intact membranes (living cells) do not absorb trypan blue, but damaged cells are permeated by trypan blue and then appear with a distinctive blue color under a microscope. Calcein-AM is a fluorescent dye with excitation and emission wavelengths of 495/515 nm. It traverses the cell membrane of living cells, making them fluoresce in green. Calcein-AM is always combined with another fluorescent dye, such as propidium iodide, in a live/dead examination to identify dead cells. Propidium iodide is normally applied as a DNA stain in microscopy to visualize the nucleus of dead cells as it cannot cross the membrane of living cells. Its fluorescent excitation and emission wavelengths are 535 and 617 nm, under which the stained cells appear red. With the combination of calcein-AM and propidium iodide, live and dead cells can be distinguished and counted under a fluorescent microscope, and thus the level of cell viability is determined.

The MTT (3-(4,5-dimethyl-2-thiazolyl)-2,5-diphenyl-2H-tetrazolium bromide) assay is a colorimetric assay for assessing cell metabolic activity and cell viability. The basic idea behind MTT is that viable cells are able to break down yellow tetrazolium into purple formazan crystals that are soluble in cell culture, thus leading to the change in color. The absorbance of the newly colored solution can be analyzed and quantified by spectrophotometry at a certain wavelength in a typical range from 500 to 600 nm. The amount of light absorption is associated with the formazan concentration accumulated in and on the cells, with a higher formazan concentration resulting in a deeper purple colour and thus greater absorbance. One limitation of the MTT assay is that it lacks the sensitivity to detect the number of viable cells; unlike live/dead cell counting, the MTT assay can only provide a rough range of cell viability. The MTT assay can also be influenced by a variety of chemical compounds that are known to interfere. The MTT reagent also exhibits cytotoxic effects that may damage or even kill cells.

Immunostaining is another technique for determining cell viability, encompassing a broad range of techniques such as immunocytochemical staining (ICC), immunofluorescent assays (IFA), and immunohistochemistry (IHC). Immunostaining uses fluorescent dyes or enzymes capable of catalyzing reactions that results in a colored product. The fluorescent dyes or colored products are easily detectable by light microscopy. In addition, radioactive elements can be used as labels, and the immunoreaction can be visualized by autoradiography, similar to positron emission tomography (PET) imaging.

## 6.7   Advanced Extrusion-Based Bioprinting Techniques

Native tissues normally contain varying cell types and extracellular matrixes, as well as other components, that are highly organized together in complex structures to perform specific functions. For the purpose of tissue regeneration, scaffolds are expected to mimic the complex structures or composition of targeted tissues and facilitate the recovery of functions [31]. As such, scaffolds produced from multiple biomaterials and cells are in high demand. For this, varying extrusion-based bioprinting techniques have been developed to print multi-materials, mainly including embedded, multi-head, coaxial, and hybrid bioprinting.

### *6.7.1   Embedded Bioprinting*

Embedded bioprinting, as shown in Fig. 6.24, is an advanced printing technique that prints bioinks into a bath containing a supporting medium to facilitate the formation and maintenance of 3D constructs. Similar to bioplotting as shown in Fig. 6.16c, embedded bioprinting also makes use of a medium, but with more requirements or functions imposed on the medium than crosslinking in bioplotting. In addition to being biocompatible with the bioink to be printed, the supporting medium in embedded bioprinting must possess the appropriate rheological properties as characterized by yield stress and viscosity. From Chap. 5, we know the yield stress of a fluid or solution is the minimum shear stress applied in order to generate the flow of a fluid or solution. In embedded bioprinting, a shear stress within the medium is generated by the needle moving inside the supporting medium that is bigger than the medium yield stress. As a result, the medium around the needle begins to flow and thus generates a void space behind the needle for bioink printing. Owing to its flow behavior, the medium then flows back and refills the space around the just-printed bioink. After that, if the shear stress within the medium generated by the effect of gravity on the bioink (while being offset by the medium buoyancy) is smaller than the medium yield stress, the medium remains stationary, thus providing mechanical support to the printed bioink and maintaining the integrity of the printed structure. As such, the supporting medium should have an appropriate value of yield

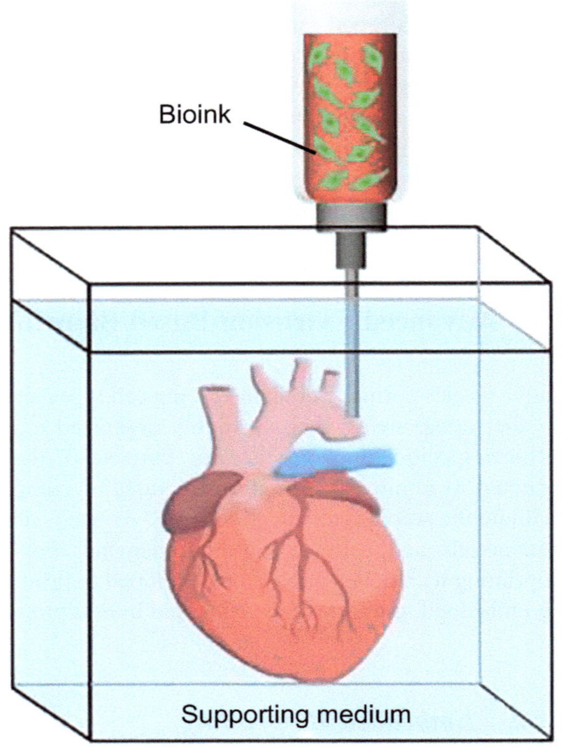



**Fig. 6.24** Schematic of embedded bioprinting, in which the bioink is printed into a supporting medium to form a complex 3D construct

stress, i.e., small enough to allow for the ready movement of the needle during printing and also large enough to support the printed bioink and constructs without movement after printing. Meanwhile, the medium viscosity should be appropriate such that it can quickly refill the space around the printed bioink. With a supporting medium with these rheological properties, embedded bioprinting can significantly enlarge the pool of bioinks that can be printed and allow for omnidirectional bioink printing to form constructs with complex structures. Common supporting media currently include carbomer, hyaluronic acid, alginate+xanthan gum, gelatin+agarose/GelMA, and laponite. In addition, the supporting medium in embedded bioprinting must be readily removed to extract the construct printed. Depending on its properties, methods to remove the supporting medium include elevating the temperature to melt the medium, enzymatic cleavage, and simple washing or dilution. The interaction between the bioink and supporting medium is also important in embedded bioprinting. In addition to biocompatibility, the bioink and supporting medium must be both chemically and physically compatible, allowing for proper crosslinking and/or minimal diffusion.

## *6.7.2  Multi-head Bioprinting*

Creating scaffolds from multiple biomaterials and cells can be achieved by pre-mixing or blending the cells in different bioinks and then utilizing multiple printing heads, with each printing one bioink similar to single-printing head bioprinting. Multiple bioinks can be individually loaded into different printing heads for scaffold printing and, by controlling the deposition from each head, multiple bioinks can be printed at different locations and organized to form complex scaffolds. One example printed from multi-head bioprinting is the hybrid scaffold comprised of a 3D framework to impart mechanical strength and a hydrogel network to incorporate living cells. Hydrogels normally impart poor mechanical support to the printed structures, which limits their applications in scaffold bioprinting. By co-printing another biomaterial such as a thermoplastic polymer as a framework to reinforce the printed hydrogel, a hybrid structure can be created that has sufficient mechanical strength. Hybrid scaffolds have been created, for example, from PCL and alginate hydrogel using multi-head bioprinting (Fig. 6.25) [32]. PCL beads were loaded into a high-temperature printing head and melted at a temperature of 65–80 ° C, while the dissolved alginate-chondrocyte suspension was loaded into a low-temperature printing head maintained at 10 °C. Based on the computer model, a PCL layer was printed first in a designed pattern, and then the alginate hydrogel was subsequently printed on the top of the PCL layer. After layer-by-layer printing, a PCL-alginate-based hybrid scaffold was created with good structural stability owing to the strong support provided by the PCL, while the alginate was able to support cell functions including cell viability, proliferation, and cartilage differentiation [32].

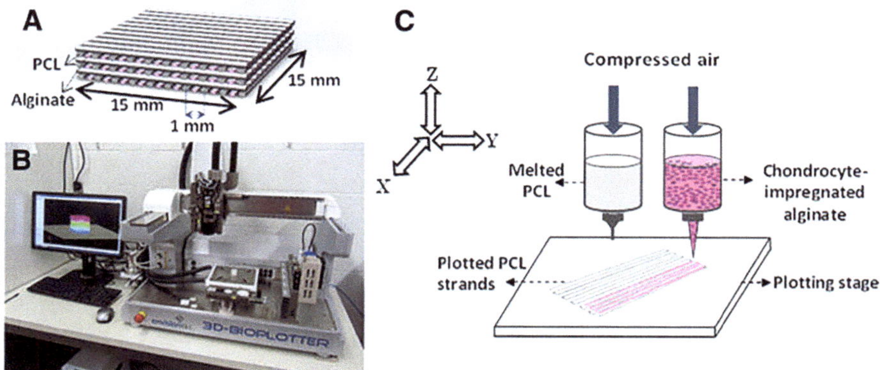


**Fig. 6.25**  Schematic of multi-head bioprinting allowing for the printing of multiple bioinks [32]

### 6.7.3   Coaxial Bioprinting

Multi-head bioprinting has the ability to produce hybrid scaffolds from multiple materials/cells, but lacks the ability to regulate or control the cross-sectional structure and/or composition of multiple materials/cells (rather than via simply blending them) in the printed strands. This ability to control the strand cross-sectional structure and/or composition is of benefit, and sometimes critical, to tissue engineering, such as with respect to vascularization as will be discussed in Chap. 7. To print such strands, coaxial bioprinting has been developed based on a coaxial needle that has a configuration featuring two (or more) dispensers with needles assembled in a coaxial configuration (Fig. 6.26); each dispenser is used to apply one bioink. As such, multiple bioinks can be extruded, respectively, through the axial dispensers and, by regulating the printing pressure applied to each dispenser, the flow rate of the bioink printed can be adjusted, thus controlling the axial-layer material distribution or portion in the cross-section of the strand. A common configuration for coaxial bioprinting is the core-shell configuration with outer and inner needles. If a biomaterial such as alginate solution is dispensed from the outer shell needle while the associated crosslinking calcium solution is printed from the inner core needle, a hollow alginate strand can be produced. If the alginate solution is dispensed through the inner needle and the calcium solution through the outer needle, the gelation of alginate inside the capillary is triggered before it is extruded, which will result in a single alginate strand with improved fidelity compared to directly printed alginate through a normal needle.

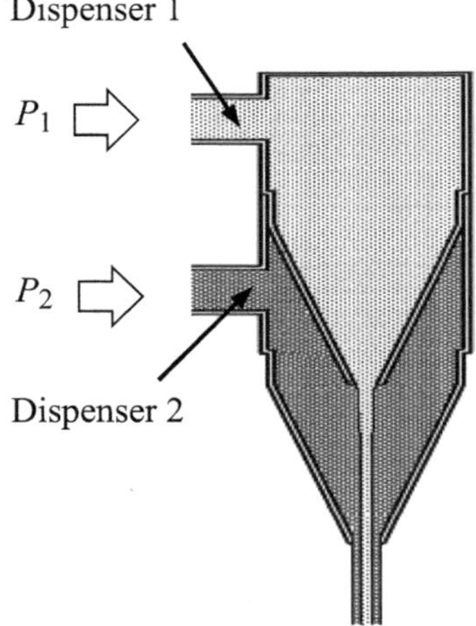


**Fig. 6.26** Coaxial bioprinting from two bioinks

### *6.7.4  Hybrid Bioprinting*

Hybrid bioprinting is an advanced method for scaffold fabrication that integrates extrusion-based bioprinting with other printing techniques. Such hybrid bioprinting capitalizes on the advantages of extrusion-based bioprinting and others techniques to create scaffolds that could not be created by means of one bioprinting technique alone. A common approach integrates extrusion-based bioprinting and an electrospinning technique with enhanced capacity to produce complex scaffolds with varying scales of strands or fibers. We know from Chap. 1 that electrospinning is a fabrication technique that can create fine fibers down to the nanometer scale (>200 nm) from polymer solutions or melts. Using a system integrating these techniques, nano- or micro-scale fibers can be produced so as to create scaffolds with varying structures and environments [33]. Such scaffolds may enhance the biological functioning, for example, by providing more cell-binding sites via nano-scale fibers to facilitate cell adhesion. Scaffolds can also be fabricated with a zonal structure, where each zone is created by one technique but with a structure and environment distinct from the next. For example, PCL nanofibers can be spun between hydrogel layers with living cells to enhance the mechanical strength of the scaffold.

## 6.8  Summary

Extrusion bioprinting works by extruding bioinks in continuous strands in a layer-by-layer manner to create constructs or scaffolds with a 3D structure. An extrusion bioprinting system mainly consists of a printing head, a three-axis positioning system, and a printing stage. The positioning system is used to move the printing head relative to the stage in three directions, while the printing head deposits the bioink onto the stage by employing a pneumatic-, piston-, or screw-driven mechanism. Due to the advantages of simple operation and ease of maintenance, pneumatic-driven bioprinting has been widely used in scaffold fabrication; however, piston- and screw-driven bioprinting allow for more direct control over the flow of bioink compared to pneumatic-driven bioprinting.

In bioprinting, the flow rate of the bioink printed is the volume (or mass) of bioink forced out of the needle per unit time; this determines the size of the printed strand and thus affects the structure of the scaffold printed. The bioink flow rate can be affected or regulated by process parameters (e.g., printing forces and temperature), structural parameters (e.g., needle geometry and size), and bioink flow behavior. In pneumatic-driven bioprinting, needle geometry is another factor affecting the bioink flow rate; under the same printing pressure, the flow rate using a tapered needle is much higher than when using a cylindrical needle. Associated with the flow rate, extrudability in bioprinting refers to the capability to extrude or print the bioink through a needle to form a continuous and controllable filament. Generally, the printing of bioink from a needle can be characterized as unextrudable, uncontinous (or jetting), continuous yet uncontrollable, or continuous and controllable. In addition to the bioink flow rate, the needle movement in the horizonal plane is important with respect to the cross-sectional profile of strands formed on the printing stage. At the SF speed, the diameter of the printed strand before its spreading is the same as the internal diameter of the needle if swelling of the material is neglected and no stresses are induced within strands. Speeds faster than the SF speed result in smaller bioink strands and induced tensile stress within the printed strand; the converse holds for slower speeds. Induced stresses (tensile or compressive) within the printed strand can further alter its molecular structure, thus affecting cell growth and functioning.

As a scaffold is a 3D structure stacked from multiple layers, the control of needle movement in the $Z$ direction is also important with respect to the stability and fidelity of the final printed scaffold. When printing the first layer, the distance between the needle tip and the printing stage should be slightly larger than the inner diameter of the needle. As more layers are printed, the distance between the needle and the printing layer is prone to increase, caused by layer deflection (or the flow or spreading of bioink) due to the influence of gravity. Strategies to compensate for layer deflection can be applied, with one effective way being to print the bioink into a medium to form scaffolds. Hydrogel bioink solutions, in particular, must be solidified or crosslinked after printing to enhance the mechanical strength and stability of the resulting printed construct. Depending on the properties of the hydrogel, various crosslinking methods can be employed (e.g., temperature control, spray, medium bath, pre-crosslinking, photopolymerization with UV light). Based on the bioink flow-rate models presented in this chapter, the bioprinting parameters of pressure and needle speed can be rigorously determined from the scaffold design as well as the flow behavior of the bioink prepared for printing, instead of trial and error methods.

Bioprinted scaffolds have porous structures to facilitate the transmission of nutrients and metabolic waste, and various techniques have been adapted or developed to characterize them in terms of pore size, pore volume/porosity, and pore interconnectivity. Generally, these techniques can be classified as non-imaging and imaging approaches. Non-imaging techniques currently include measuring the strand profiles, gravimetry, and mercury porosimetry; imaging techniques include 3D micro-computed tomography ($\mu$CT) and 2D scanning electron microscopy (SEM). Compared to non-imaging techniques, $\mu$CT and SEM have become more popular for characterizing porous scaffolds as they can provide more information related to scaffold morphology and structure.

Cell printing demands the use of hydrogels that provide a compatible, aqueous environment for incorporated cells. Cells can be supported by the highly hydrated hydrogel network that permits the exchange of nutrients from the culture environment and transport of wastes produced by the cells. Maintaining cell viability and other cell functions during cell printing is very important. Process-induced cell damage during bioprinting can lead to a reduced number of living cells for subsequent functions and tissue regeneration. Process-induced mechanical forces contribute to cell damage and include compressive pressure, shear stress, and extensional stress. Cells are subjected to these forces as they are forced to flow through the bioprinting syringe and needle. Characterizing cell damage by considering the relationship between these forces and cell damage is crucial to understand the mechanism of cell damage in the bioprinting process, which will inform new printing strategies that reduce cell damage and preserve cell viability. Techniques such as microscopy combined with different dyes, MTT assays, and immunostaining are often used to measure cell viability in bioprinting.

Bioprinting scaffolds with multiple materials and cells has attracted considerable attention related to efforts to mimic native tissue components. Various extrusion-based bioprinting techniques have been developed and advanced, such as embedded bioprinting, multi-head bioprinting, coaxial bioprinting, and hybrid bioprinting. More advanced extrusion-based bioprinting techniques are expected to be developed in the future for tissue-like scaffold fabrication.

**Problems**

1. How can extrusion-based bioprinting systems be classified based on the printing mechanisms? What are the advantages and disadvantages of each system?
2. A 2% w/v bioink is prepared with a measured density of 1 g/mL. Its flow behavior is measured using a cone-and-plate rheometer, with the shear stress and shear rate recorded as given below.

| Shear rate ($s^{-1}$) | Shear stress (Pa) |
| --- | --- |
| 0.15 | 0.02 |
| 1 | 0.718 |
| 10 | 13.593 |
| 100 | 78.34 |
| 200 | 155.83 |
| 400 | 290.43 |
| 800 | 533.92 |
| 1600 | 834.67 |

A 300-μm diameter, 12-mm long cylindrical needle is used for scaffold printing. During the printing procedure, the extrusion pressure is set at 100 kPa.

(a) If the bioink can be expressed by a power-law model, identify the indexes $K$ and $n$.
(b) If the flow inside the needle is fully developed, there is no wall slip, and the flow behavior is independent of temperature, calculate the volumetric flow rate of the bioink as well as the velocity at the central line of the needle.
(c) Suppose the printed bioink does not spread and the strands have a cylindrical geometry after printing. Calculate the strand diameter if the dispenser speed is 10 mm/s.

3. Examine the bioprinting processes illustrated below and explain the reasons behind the varying diameter and geometries of strands formed. Suppose the printed bioink owns non-Newtonian flow property that can be expressed by a power-law model.

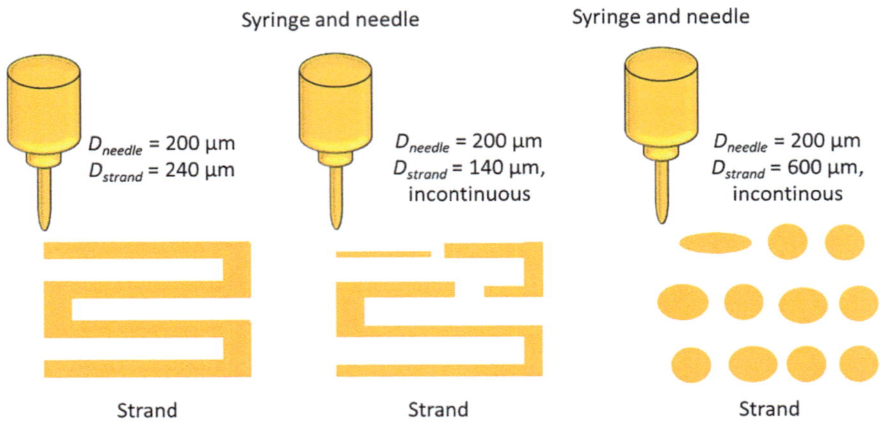


4. A scaffold design has a strand diameter of 250 µm. Suppose the scaffold is to be printed from a biomaterial solution with the flow behavior described by $\tau = 15\dot{\gamma}^{0.5}$, illustrate how you select the printing conditions including the needle gauge (with a cylindrical shape), applied air pressure, and needle speed.

5. Briefly explain the needle geometries (focusing on cylindrical vs. tapered needles) on the bioink flow rate and cell damage.

6. Name two of the cases related to the extrudability in bioprinting and explain the reasons or mechanisms behind.

7. Briefly explain the effects of needle movement in both horizonal plane and vertical direction on the strand profiles and scaffold structures.

8. Identify a study from the bioprinting literature to briefly explain one crosslinking technique used in bioprinting.

9. Name and describe two techniques to characterize the strand profile and scaffolds structures.

10. Name and briefly explain the bioprinting process-induced mechanical forces, and explain their effects on the cell deformation and damage.

11. Identify a study from the recent literature to briefly explain how the cell viability in bioprinting is measured and characterized.

12. A cell/hydrogel bioink is printed by means of a 200-µm diameter, 20-mm long cylindrical needle from an extrusion-based bioprinting system at room temperature. The printing pressure is 200 kPa and the flow inside the needle is fully developed.

(a) The diameter of the printing syringe is known to be much greater than the needle diameter. Determine the hydrostatic pressures experienced by the cells located at different positions in the printing system as shown in the figure below (do not consider the pressure drop due to the contractive region).

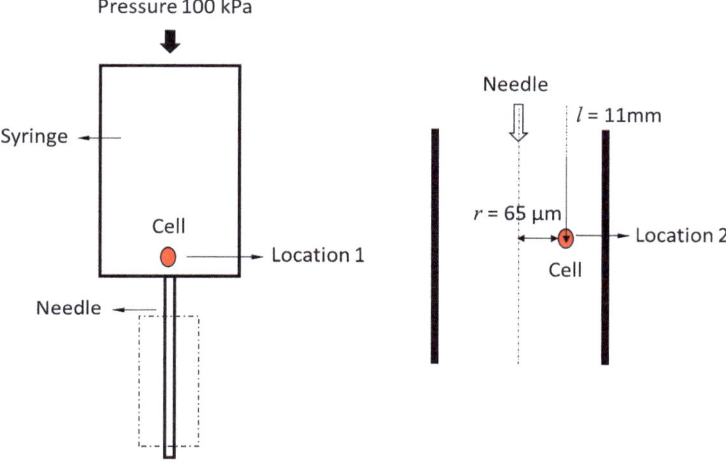


(b) Determine the shear stress experienced by the cells inside the needle (location 2), and calculate the time required for a cell to travel through the whole needle tip if it follows the vertical streamline from the inlet to the outlet (suppose the power-law indexes $K$ and $n$ are 31.2 and 0.33, respectively).

13. Name and briefly describe two advanced extrusion-based bioprinting techniques, along with their advances or merits.

# References


1. L. Ning, X. Chen, A brief review of extrusion-based tissue scaffold bio-printing. Biotechnol. J. **12**, 1600671 (2017). https://doi.org/10.1002/biot.201600671
2. X.B. Chen, G. Schoenau, W.J. Zhang, On the flow rate dynamics in time-pressure dispensing processes. J. Dyn. Syst. Meas. Control. **124**, 693–698 (2002). https://doi.org/10.1115/1.1514058
3. X. Chen, Dispensed-based bio-manufacturing scaffolds for tissue engineering applications. Int. J. Eng. Appl. **2**, 9–10 (2014)
4. X.B. Chen, M.G. Li, H. Ke, Modeling of the flow rate in the dispensing-based process for fabricating tissue scaffolds. J. Manuf. Sci. Eng. **130**(2), 021003 (2008). https://doi.org/10.1115/1.2789725

5. M. Sarker, X.B. Chen, Modeling the flow behavior and flow rate of medium viscosity alginate for scaffold fabrication with a three-dimensional bioplotter. J. Manuf. Sci. Eng. **139**(8), 081002 (2017). https://doi.org/10.1115/1.4036226

6. M.G. Li, X.Y. Tian, X.B. Chen, Modeling of flow rate, pore size, and porosity for the dispensing-based tissue scaffolds fabrication. J. Manuf. Sci. Eng. **131**(3), 034501 (2009). https://doi.org/10.1115/1.3123331

7. M. Li, X. Tian, D.J. Schreyer, X. Chen, Effect of needle geometry on flow rate and cell damage in the dispensing-based biofabrication process. Biotechnol. Prog. **27**, 1777–1784 (2011). https://doi.org/10.1002/btpr.679

8. X.B. Chen, H. Ke, Effects of fluid properties on dispensing processes for electronics packaging. IEEE Trans. Compon. Packag. Manuf. Technol. **29**, 75–82 (2006). https://doi.org/10.1109/TEPM.2006.874964

9. X.B. Chen, Modeling of rotary screw fluid dispensing processes. J. Electron. Packag. **129**, 172–178 (2006). https://doi.org/10.1115/1.2721090

10. X.B. Chen, J. Kai, Modeling of positive-displacement fluid dispensing processes. IEEE Trans. Compon. Packag. Manuf. Technol. **27**, 157–163 (2004). https://doi.org/10.1109/TEPM.2004.843083

11. Z. Fu, S. Naghieh, C. Xu, et al., Printability in extrusion bioprinting. Biofabrication **13**, 033001 (2021). https://doi.org/10.1088/1758-5090/abe7ab

12. S. England, A. Rajaram, D.J. Schreyer, X. Chen, Bioprinted fibrin-factor XIII-hyaluronate hydrogel scaffolds with encapsulated Schwann cells and their in vitro characterization for use in nerve regeneration. Bioprinting **5**, 1–9 (2017). https://doi.org/10.1016/j.bprint.2016.12.001

13. L. Ning, N. Zhu, F. Mohabatpour, et al., Bioprinting Schwann cell-laden scaffolds from low-viscosity hydrogel compositions. J. Mater. Chem. B **7**, 4538–4551 (2019). https://doi.org/10.1039/C9TB00669A

14. Y. Jin, D. Zhao, Y. Huang, Study of extrudability and standoff distance effect during nanoclay-enabled direct printing. Bio-Des. Manuf. **1**, 123–134 (2018). https://doi.org/10.1007/s42242018-0009-y

15. A. GhavamiNejad, N. Ashammakhi, X.Y. Wu, A. Khademhosseini, Crosslinking strategies for 3D bioprinting of polymeric hydrogels. Small **16**, 2002931 (2020). https://doi.org/10.1002/smll.202002931

16. A. Rajaram, D. Schreyer, D. Chen, Bioplotting alginate/hyaluronic acid hydrogel scaffolds with structural integrity and preserved Schwann cell viability. 3D Print Addit. Manuf. **1**, 194–203 (2014). https://doi.org/10.1089/3dp.2014.0006

17. N. Zhu, M.G. Li, D. Cooper, X.B. Chen, Development of novel hybrid poly(l-lactide)/chitosan scaffolds using the rapid freeze prototyping technique. Biofabrication **3**, 034105 (2011). https://doi.org/10.1088/1758-5082/3/3/034105

18. X. Duan, N. Li, X. Chen, N. Zhu, Characterization of tissue scaffolds using synchrotron radiation microcomputed tomography imaging. Tissue Eng. Part C Methods **27**, 573–588 (2021). https://doi.org/10.1089/ten.tec.2021.0155

19. L. Ning, N. Zhu, A. Smith, et al., Noninvasive three-dimensional in situ and in vivo characterization of bioprinted hydrogel scaffolds using the x-ray propagation-based imaging technique. ACS Appl. Mater. Interfaces **13**, 25611–25623 (2021). https://doi.org/10.1021/acsami.1c02297

20. L. Ning, H. Sun, T. Lelong, et al., 3D bioprinting of scaffolds with living Schwann cells for potential nerve tissue engineering applications. Biofabrication **10**, 035014 (2018). https://doi.org/10.1088/1758-5090/aacd30

21. A. Sadeghianmaryan, S. Naghieh, H. Alizadeh Sardroud, et al., Extrusion-based printing of chitosan scaffolds and their in vitro characterization for cartilage tissue engineering. Int. J. Biol. Macromol. **164**, 3179–3192 (2020). https://doi.org/10.1016/j.ijbiomac.2020.08.180

22. M. Li, X. Tian, N. Zhu, et al., Modeling process-induced cell damage in the biodispensing process. Tissue Eng. Part C Methods **16**, 533–542 (2010). https://doi.org/10.1089/ten.tec.2009.0178

23. L. Ning, A. Guillemot, J. Zhao, et al., Influence of flow behavior of alginate–cell suspensions on cell viability and proliferation. Tissue Eng. Part C Methods **22**, 652–662 (2016). https://doi.org/10.1089/ten.tec.2016.0011
24. L. Ning, B. Yang, F. Mohabatpour, et al., Process-induced cell damage: Pneumatic versus screw-driven bioprinting. Biofabrication **12**, 025011 (2020). https://doi.org/10.1088/1758-5090/ab5f53
25. N. Soltan, L. Ning, F. Mohabatpour, et al., Printability and cell viability in bioprinting alginate dialdehyde-gelatin scaffolds. ACS Biomater Sci. Eng. **5**, 2976–2987 (2019). https://doi.org/10.1021/acsbiomaterials.9b00167
26. L. Ning, N. Betancourt, D.J. Schreyer, X. Chen, Characterization of cell damage and proliferative ability during and after bioprinting. ACS Biomater Sci. Eng. **4**, 3906–3918 (2018). https://doi.org/10.1021/acsbiomaterials.8b00714
27. F.N. Cogswell, Measuring the extensional rheology of polymer melts. Trans. Soc. Rheol. **16**, 383–403 (1972). https://doi.org/10.1122/1.549257
28. D.M. Binding, An approximate analysis for contraction and converging flows. J. Non-Newton Fluid Mech. **27**, 173–189 (1988). https://doi.org/10.1016/0377-0257(88)85012-2
29. K. Nair, M. Gandhi, S. Khalil, et al., Characterization of cell viability during bioprinting processes. Biotechnol. J. **4**, 1168–1177 (2009). https://doi.org/10.1002/biot.200900004
30. M.G. Li, X.Y. Tian, X. Chen, Temperature effect on the shear-induced cell damage in biofabrication. Artif. Organs **35**, 741–746 (2011). https://doi.org/10.1111/j.1525-1594.2010.01193.x
31. N. Betancourt, X. Chen, Review of extrusion-based multi-material bioprinting processes. Bioprinting **25**, e00189 (2022). https://doi.org/10.1016/j.bprint.2021.e00189
32. Z. Izadifar, T. Chang, W. Kulyk, et al., Analyzing biological performance of 3D-printed, cell-impregnated hybrid constructs for cartilage tissue engineering. Tissue Eng. Part C Methods **22**, 173–188 (2016). https://doi.org/10.1089/ten.tec.2015.0307
33. X.B. Chen, A. Fazel Anvari-Yazdi, X. Duan, et al., Biomaterials/bioinks and extrusion bioprinting. Bioact. Mater. **28**, 511–536 (2023). https://doi.org/10.1016/j.bioactmat.2023.06.006

# Chapter 7
# Bioprinting of Vascular Networks in Scaffolds

## 7.1 Introduction

An interconnected vascular network is required to maintain the viability and biological function of large cell populations in growing tissue. *In vivo*, well-distributed vascular capillaries are seen in different tissues at a distance of every ~100–200μm. Similarly, tissue regeneration with the aid of scaffolds, particularly large and thick scaffolds, requires the incorporation of an interconnected vascular network to facilitate mass transfer of nutrients, signaling molecules, oxygen, growth factors, metabolic waste, etc., between the cells in scaffolds and the culture medium. Understanding *in vivo* blood vessels and their formation is essential to the design and fabrication of functional vascular networks within engineered scaffolds. This first part of this chapter discusses the vascular anatomy and capillary vessel formation mechanisms *in vivo*. To fabricate vascularized scaffolds, direct and indirect approaches based on bioprinting have been developed to create capillary-like structures or macro blood vessels. The direct approach allows the biofabrication of lumen-containing strands within the scaffolds, analogous to native vessels. In this approach, cell/hydrogel mixtures are used as a bioink, with growth factors often added to enhance the biofunctionality of the bioink. In the indirect approach, vascular networks are generated within the scaffold by removing sacrificial strands that are created by bioprinting or other additive manufacturing techniques. Biopolymer free approaches, such as decellularized native tissue, self-assembled tissue filaments or cell aggregates, and cell sheets, have also been considered for the formation of macro or micro blood vessels. The second part of this chapter introduces these bioprinting-based approaches to create vascular networks within tissue scaffolds, along with the merits and limitations associated with each approach.


D. X. B. Chen, *Extrusion Bioprinting of Scaffolds for Tissue Engineering*, https://doi.org/10.1007/978-3-031-72471-8_7

## 7.2   Blood Vessels and Formation Mechanisms

The vascular network is comprised of arteries and veins, where arteries supply oxygenated blood to tissues and veins take carbon dioxide ($CO_2$)-enriched blood away from tissues (Fig. 7.1). *In vivo*, major arteries branch out into arterioles, metarterioles, and arterial capillaries. Arteries are composed of five sequential layers of adventitia, elastic lamina, smooth muscle cells (SMCs), basement membrane (BM), and endothelial cells (ECs); arterioles of three successive layers of SMCs, BM, and ECs; and capillaries of two consecutive layers of BM and ECs. Numerous capillaries combine downstream to form post-capillary venules, and then bunches of venules combine to create major veins. In the vascular network, arteries supply blood to all tissues and organs from the heart, arterioles contract and expand to control blood flow, and capillaries facilitate mass transfer (e.g., nutrients, oxygen, $CO_2$, uric acid, water) between blood and the surrounding interstitial fluid. In contrast, venules collect the returned blood from the capillary bed and drain the deoxygenated blood to larger veins. In the capillary bed, the precapillary sphincter regulates the blood flow while the thoroughfare channel maintains continuous blood flow.

*In vivo*, continuous, fenestrated, and sinusoidal capillaries are mainly responsible for diffusional mass transfer (Fig. 7.2). Most capillaries within the body are continuous and exchange mass (water, gas molecules, ions, and other water-soluble molecules) through diffusion, vesicles, pinocytosis, and intercellular clefts. Endocrine glands, the intestines, pancreas, and kidneys contain nano-porous (60–80 nm) fenestrated capillaries, where mass transfer occurs through intercellular clefts, pores, diffusion, and vesicles. Microporous (30–40μm) sinusoidal capillaries are found in the adrenal gland, liver, spleen, and bone marrow [1]. Sinusoidal capillaries, which are leaky in nature, allow the transport of white and red blood cells as well as serum protein.

Angiogenesis and vasculogenesis are the two mechanisms by which blood vessels form in the womb and after birth. Angiogenesis is the process by which new micro-vessels sprout from existing blood vessels after being stimulated by

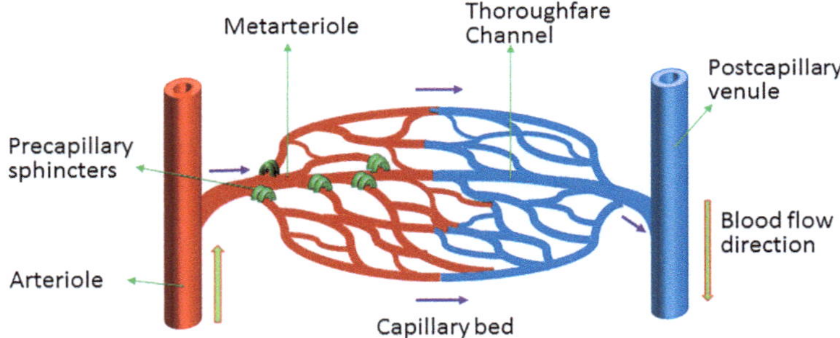


**Fig. 7.1** Schematic of *in vivo* blood circulatory system

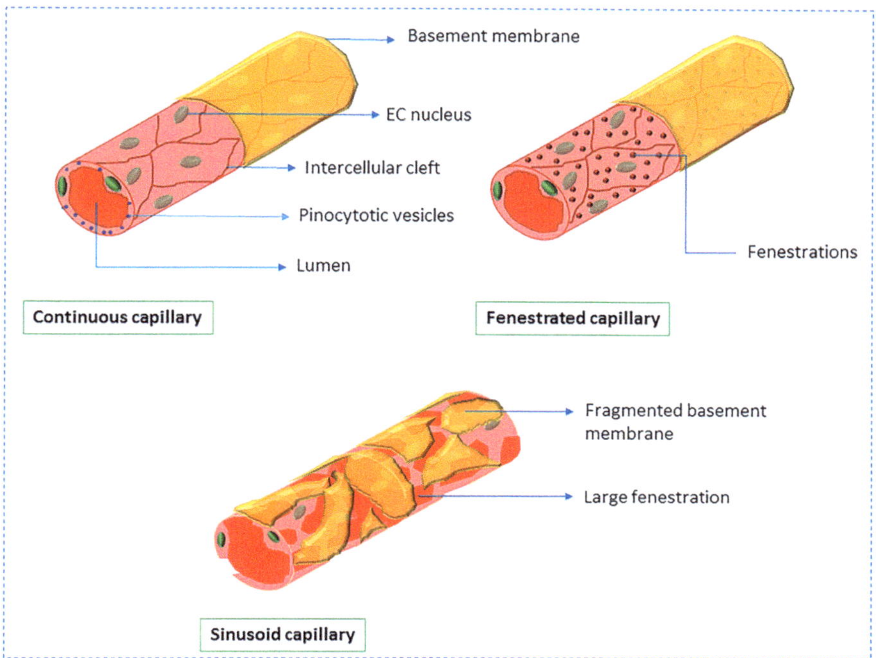



**Fig. 7.2** Schematic of different types of micro blood capillaries found in tissues and organs

various angiogenic factors and biochemical signals. The sprouting blood vessels grow into ischemic tissue in response to attractive or repulsive biochemical signals, mechanical cues, and gradients in the tissue. Random sprouting of capillaries results in the development of an immature and excessive vascular plexus, which reduces the efficiency of the vascular network. Therefore, the vascular plexus is frequently remodeled to meet the demands of the tissue and achieve a functional and mature vascular network. The intricate process of angiogenesis is illustrated in a few steps in Fig. 7.3. Generally, vascular endothelial growth factor (VEGF) secreted by ischemic tissue dilates the existing blood vessel, increases vascular permeability, and facilitates EC migration. Angiopoietin (Ang-2) promotes angiogenesis by degrading the ECM protein and eliminating SMCs from capillaries [2]. Interestingly, microvascular ECs also release various matrix metalloproteases (such as MMP-2, MMP-3, MMP-9) and tissue inhibitor of metalloproteinase-2 (TIMP2) to regulate basement membrane and ECM degradation to promote EC migration [3]. In addition, the proteolytic degradation of ECM releases several growth factors including insulin-like growth factor-1 (IGF-1), VEGF, and basic fibroblast growth factor (FGF-2), which play a significant role in angiogenic sprouting [4]. The progression process starts after the degradation of basement membrane and ECM. In this process, ECs migrate and proliferate into ECM by the regulation of VEGF, platelet-derived growth factor (PDGF), FGFs, IGF-1, neuropeptides, erythropoietin, angiopoietins,

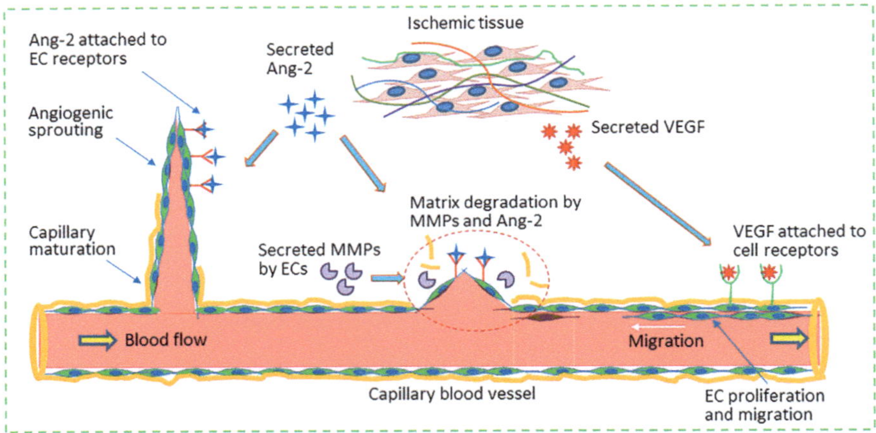


**Fig. 7.3**  Schematic of blood capillary formation by angiogenesis

interleukins, and hepatocytes [5, 6]. Thereafter, differentiated ECs generate a cord-like structure, where lumen starts forming due to the combined effect of Ang-2 and VEGF. Because $VEGF_{121}$, $VEGF_{165}$, and Ang-1 enhance lumen diameter and $VEGF_{189}$ reduces lumen diameter, the balance among the different angiogenic factors controls the lumen diameter within the regenerated capillaries [7].

In vasculogenesis, bone marrow- or blood-derived endothelial progenitor cells (EPCs) play a significant role in vasculature formation through a complex process (Fig. 7.4). Due to exogenous or endogenous influences, secreted growth factors, cytokines, and hormones promote the proliferation, differentiation, and mobilization of EPCs in the bone marrow. In particular, a group of chemo-attractant factors, such as basic fibroblast growth factor (bFGF), stromal cell-derived factor-1 (SDF-1), placental growth factor, erythropoietin, granulocyte colony stimulating factor, granulocyte–monocyte colony stimulating factor, and VEGF, regulate the movement of EPCs from bone marrow and the recruitment at the unvascularized or ischemic site [8]. The released EPCs circulate in the peripheral blood and settle at the ischemic tissue site through chemotaxis, adhesion, and transendothelial migration. Gradients of chemokines (SDF-1, interleukin-8 (IL-8), growth-regulated onco-gene-α, and C–C chemokine) regulate the chemotaxis and adhesion of EPCs to the inner layer of blood vessels [9]. After transendothelial migration, EPCs reach the ischemic tissue site by rupturing the basement membrane and ECM using extracellular proteases (MMP-9, cathepsin L, urokinase-type plasminogen activator, and tissue-type plasminogen activator) [10]. Upon accumulation at the ischemic site, EPCs form a vascular pattern through proliferation, differentiation, and interaction with existing ECs and ECM. In this process, immunoglobulin, epidermal growth factor (EGF), and VEGF promote proliferation [11], while IGF-1, monocyte chemoattractant protein-1, VEGF, PDGF, and SDF-1 upregulate the differentiation of EPCs [12].

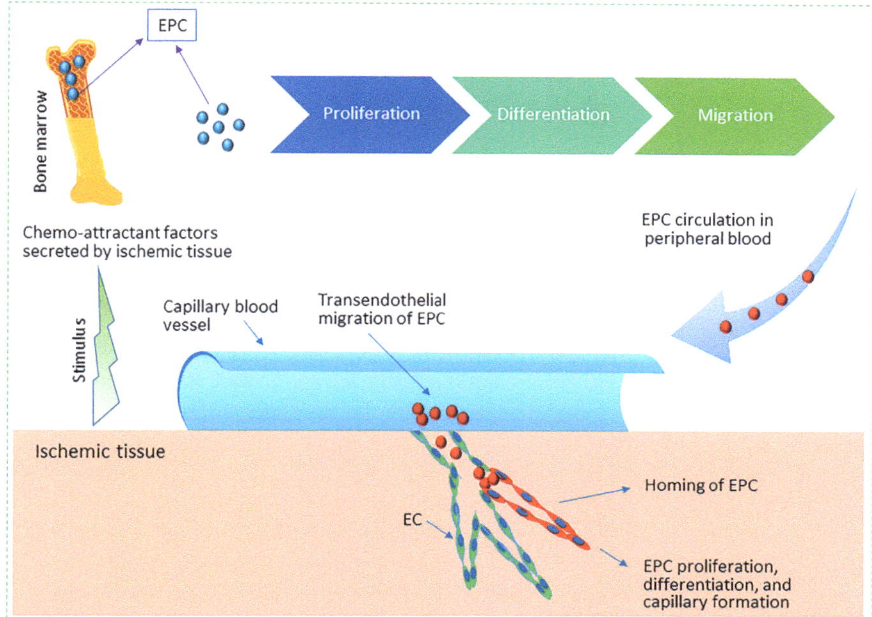


**Fig. 7.4** Schematic of blood capillary formation by vasculogenesis

## 7.3   Bioinks for Vascular Networks

Bioinks for vascularized scaffolds typically consist of biomaterials, vascular cells, and/or angiogenic factors. As the supporting component of a bioink, biomaterials must facilitate the development of functional vasculature by supporting the survival, proliferation, differentiation, and migration of vascular cells, in addition to being printable, mechanically stable, and biodegradable. Biomaterials used for printing vascular networks, similar to most bioinks, are polymers, either natural or synthetic as discussed in Chap. 3. The natural polymers for vascular networks are either protein-based or polysaccharidic. Common protein-based polymers include fibronectin, fibrin, elastin, silk, fibroin, Matrigel, and collagen, while common polysaccharidic polymers include alginate, chitosan, and hyaluronic acid [13]. Synthetic polymers have also been employed in vascularization studies due to their superb mechanical properties, biodegradability, biocompatibility, and non-toxicity, though they lack bioactivity/bioactive molecules. Common synthetic polymers include polyglycolic acid (PGA), polylactic acid (PLA), polycaprolactone (PCL), and their composites and derivatives.

Vascular cells are those seen in the blood *in vivo*, and mainly include ECs, smooth muscle cells (SMCs), and pericytes. These primary cells can be harvested from autologus, xenogenic, or allogenic sources. Because the use of xenogenic or allogenic vascular cells may provoke host immune response, autologous vascular cells are a better choice to avoid immunological rejection. Nowadays, vascular cells, harvested from sources such as the umbilical vein, dermal microvessels, omentum fat, and carotid arteries, have been frequently investigated in both angiogenesis and vasculogenesis studies [13]. Varying vascular cells are incorporated into scaffolds to promote vascular networks, though the ideal cell types and/or their co-culture combinations are still to be discovered. Instead of primary cells, stem cells can also be incorporated into vascular scaffolds, serving as renewable cellular sources to form vascular networks. Mesenchymal stem cells (MSCs), for example, obtained from bone marrow, adipose tissue, blood, or dermis have the potential to differentiate into perivascular cells (e.g., SMCs) *in vivo*.

Angiogenic factors play a significant role in forming a vascular network within engineered scaffolds through either angiogenesis or vasculogenesis. Vasculature can be formed within engineered scaffolds through the interplay of multiple angiogenic factors, typically including VEGF, bFGF, PDGF, Ang, TGF-$\beta$, sphingosine-1-phosphate, and hepatocyte growth factor (HGF). For example, it has been illustrated that co-delivery of VEGF/bFGF followed by the release of PDGF (a sequential GF release) is more effective for promoting angiogenesis than the delivery of VEGF alone or in combination with bFGF [14]. Furthermore, spatiotemporal release of GFs appears essential to guide angiogenesis in engineered scaffolds; however, there are still many challenges to overcome. Angiogenic factors can be directly loaded into bioinks or incorporated into bioinks by means of micro- or nano-particles as discussed in Chap. 8. Sustained release of angiogenic factors can also be achieved by gene-transfected cells [13].

## 7.4  Bioprinting Vascular Networks

### *7.4.1  Direct Bioprinting*

In direct bioprinting of vascular networks, the hydrogel is printed in the form of strands to form scaffolds, with ECs incorporated into the bioink prior to printing, or seeded and/or migrated after printing, to form capillary blood vessels. Culturing scaffolds with so-printed vascular networks in a bioreactor along with appropriate physiologic conditions (i.e., pulsatile flow) results in the formation of micro-capillaries. Notably, the vascular network formed in this way might not be well connected, perfused, or functional mainly due to an insufficient cell population, loss of EC attachment to the strands, and migration of ECs.

Coaxial bioprinting has been explored to tackle this issue. Direct bioprinting with coaxial needles allows the fabrication of lumen-containing strands, also known as shell-core filaments. In the coaxial system shown in Fig. 7.5b, two or three needles with different opening diameters are assembled coaxially while separate extrusion arrangements are used to control the flow of the biopolymer, cell suspension, and ionic crosslinker. In forming lumen-containing strands, ECs mixed with polymer are

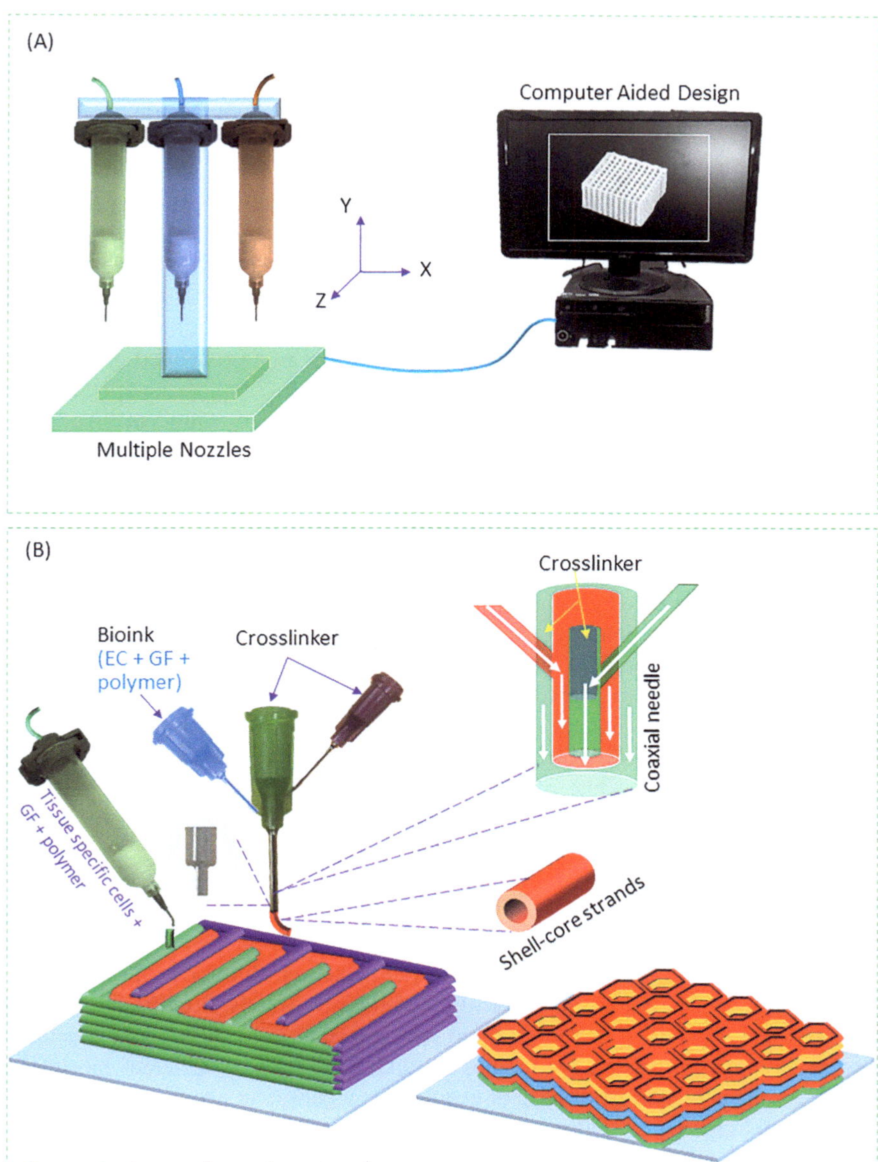


**Fig. 7.5** Schematic of direct bioprinting of vascularized scaffold: (**a**) multiple nozzle system and (**b**) bioprinting capillary blood vessel with a coaxial needle

extruded as the shell, while ionic crosslinker is dispensed as the core. To further crosslink the shell, the strands are often dispensed into ionic crosslinker, which might reduce the viability of the incorporated cell population due to long exposure to the crosslinker. The outer shell of the strand can simultaneously be crosslinked with the inner core if crosslinkers are dispensed in both the core and outer shell from a three-needle coaxial nozzle.

In a coaxial bioprinting system, lumen-containing strands are extruded in a layer-by-layer fashion as per a pre-drawn structure developed in a CAD program. The virtual 3D structure of the vascular network created in the CAD software is segmented into a number of 2D slices, and the coaxial bioprinting system prints each successive 2D slice to complete the 3D capillary bed. In particular, the system moves its coaxial needle in the x, y, and z directions to bioprint a predefined 3D vascular network. With the help of a multiple nozzle system (Fig. 7.5a), tissue-specific cells are extruded along with shell-core strands side by side to mimic *in vivo* tissue formation.

Hydrogel-based bioinks are often used in coaxial fabrication systems due to their hydrated nature and tissue-like properties. However, the CF system does not allow all concentrations of hydrogel precursors to be printed. In most cases, low concentration hydrogels have low viscosity and high concentration hydrogels have high viscosity. Extrusion of very low viscosity hydrogel precursors from a coaxial needle causes fabrication complexity while high viscosity polymers result in reduced viability of the incorporated ECs. The printed strands should be geometrically identical to the virtual structure developed in the CAD program; however, the actual printed structure often deviates from the predefined dimension based on the bioink printability. The printability of hydrogel-based bioinks depends on the viscosity of the biopolymer, gelation rate of the ionic crosslinker, and the hydrogel precursor [15]. Bioink dispensed from a 3D bioprinter needs to be crosslinked quickly to maintain acceptable printability.

The flow rate of bioink and crosslinker is critical in a coaxial biofabrication system. Arbitrary selection of bioink and crosslinker feed rates is not conducive to successful biofabrication, as random selection might interrupt the flow of bioink and clog the needle. In particular, the feed rate and concentration of bioink and crosslinker affect the lumen diameter of the shell-core strands. For different bioinks and ionic crosslinkers, the feed rate and concentration must be optimized prior to biofabrication. A number of outcomes are likely by varying different fabrication variables during coaxial printing. For example, at a particular needle speed and feed rate of bioink, increased hydrogel precursor concentrations reduce the lumen diameter of the printed strands. Further, simultaneous increase of the feed rate of the bioink and crosslinker at a particular concentration decreases the lumen diameter. Conversely, lumen diameter increases with both elevated feed rate of crosslinker for a particular feed rate of bioink or elevated feed rate of bioink at a specific feed rate of crosslinker in the coaxial needle. Moreover, for a particular feed rate, an elevated concentration of ionic crosslinker or needle speed reduces the lumen diameter of shell-core strands [16].

Within the shell-core strands, the mechanical properties and biological performance of the biopolymer significantly influence the viability, proliferation, differentiation, and migration of ECs that regulate the formation of blood vessels. A good strategy is to add peptide or protein molecules in the biopolymer shell if it lacks cell binding sites. Incorporation of growth factors (e.g., VEGF, PDGF) significantly affects the development of capillaries. However, short half-lives of growth factors are a major shortcoming that limits their application. Direct fabrication requires a large EC population to be incorporated in the shell. Biopsy is an established procedure to harvest primary ECs from the patient; however, ECs obtained in this way demonstrate inadequate proliferation and differentiation. In contrast, ECs obtained from controlled differentiation of stem cells using transcription factors show outstanding proliferation and differentiation.

**Example 7.1**
You need to fabricate a $10 \times 10 \times 10 \text{ mm}^3$ vascular network with a coaxial needle as per the following specifications:

(a) Radius of shell $= 250\mu\text{m}$.
(b) Radius of core $= 100\mu\text{m}$.
(c) Number of strands in each layer $= 5$.
(d) 4% (w/v) hydrogel precursor solution (solvent was water)
(e) Cell density $= 10^6$ cells/mL hydrogel precursor.

   The volume of a coaxial strand is

$$V = \pi L \left( R_S^2 - R_C^2 \right),$$

where $L$, $R_S$, and $R_C$ are the strand length, strand radius of the shell, and strand radius of the core, respectively.

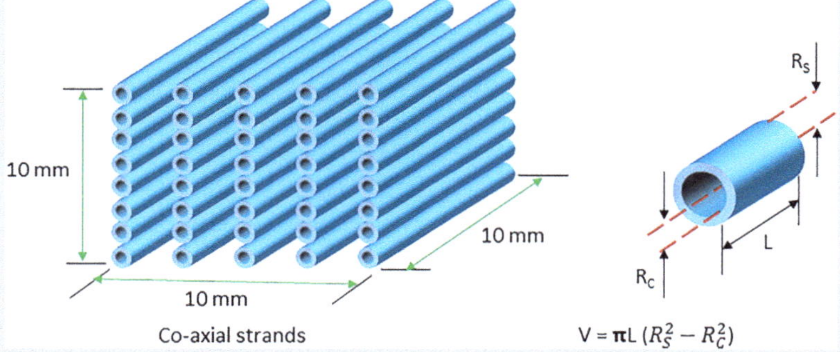


   Estimate the required amount of biopolymer and cells to fabricate the scaffold.

**Solution**
Total number of 2D layers $= (10 \text{ mm} \div 500\mu\text{m}) = (10 \text{ mm} \div 0.5 \text{ mm}) = 20$.

Number of strands in each layer $= 5$.

Total length of strands in each layer $= (5 \times 10) \text{ mm} = 50 \text{ mm}$.

Total length of strands in whole scaffold $= (20 \times 50) = 1000 \text{ mm}$.

Required volume of hydrogel precursor solution $= \pi L \ (R_S^2 - R_C^2) = (3.14)$ $(1000 \text{ mm}) (0.25^2 - 0.1^2) \text{ mm}^2 = 164.85 \text{ mm}^3 = 0.165 \text{ mL}$.

4% (w/v) hydrogel precursor solution contains 4 g of biopolymer in 100 mL solution.

Total amount of biopolymer $= (0.165 \text{ mL hydrogel precursor solution}) (4 \text{ g biopolymer}/100 \text{ mL hydrogel precursor solution}) = 0.007 \text{ g}$ **(Ans.)**

Required cells $= (0.165 \text{ mL hydrogel precursor solution}) (10^6 \text{ cells/mL hydrogel precursor solution}) = 1.65 \times 10^5 \text{ cells}$ **(Ans.)**


## 7.4.2 *Indirect Bioprinting*

Incorporation of sacrificial strands into 3D scaffolds is an indirect approach to create vascular networks within the hydrogel. The 3D printer or bioplotter allows for the fabrication of sacrificial strands (from fugitive ink, alginate, Pluronic® F127, or carbohydrate glass) layer by layer to obtain complex 3D vascular networks (Fig. 7.6a). To encapsulate such vascular networks, hydrogels or cell/peptide-loaded hydrogels (e.g., poly(ethylene glycol), gelatin methacrylate) are used and crosslinked (Fig. 7.6b). Upon crosslinking, the sacrificial filaments are removed using appropriate solvents (e.g., ethylenediaminetetraacetic acid (EDTA), Dulbecco's Modified Eagle's Medium (DMEM)), temperature, or pressure, which results in the formation of interconnected vascular channels within the hydrogel. After removal of the filaments and successive washing, ECs are generally seeded on the wall of the capillary lumen, where the cells attach, proliferate, and differentiate to form microcapillaries. Ordinary injection or perfusion bioreactors are used to seed ECs, where the cell suspension is fed through the interconnected vascular network (Fig. 7.6c). The tissue construct is cultured in an incubator for a given time period while pulsatile flow of culture media through the vascular network is maintained by a bioreactor. Perfusion with culture media under pulsatile flow influences the proliferation and differentiation of ECs to form vasculature.

Strands printed using ordinary fugitive ink (i.e., mixture of wax, oil, and nanoparticles) deform in a large tissue construct, while microcrystalline wax-based fugitive ink maintains its shape in the fabrication process and facilitates the formation of an interconnected capillary network within the scaffold after removal. However, the wax content in fugitive ink has limits with respect to reasonable printability; fugitive ink with an elevated wax content (>50 wt.%) demonstrates poor printability and ruptured strands while lower wax content shows outstanding printability [17]. Another fugitive ink, Pluronic F127, is attractive in the fabrication

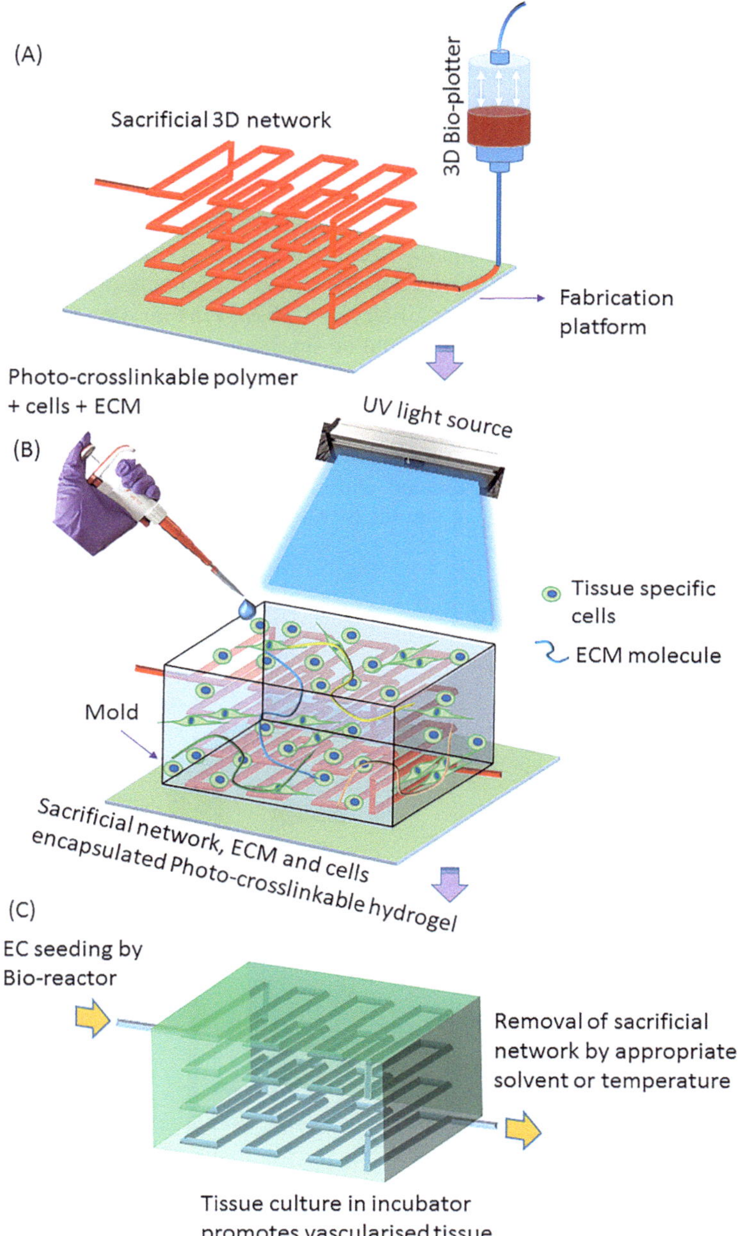


**Fig. 7.6** Indirect fabrication approach to create a vascular network within a 3D cell-incorporated scaffold: (**a**) printing of a 3D sacrificial network of strands, (**b**) encapsulation of sacrificial network, (**c**) removal of sacrificial network and seeding of ECs on the capillary lumen

of sacrificial networks due to outstanding printability and easy removal under mild conditions [18]. Moreover, cytocompatible sacrificial template carbohydrate glass is a potential alternative to fugitive ink with respect to improving cell viability. Carbohydrate glass, a composite of glucose, sucrose, and dextran, requires a thin coating of poly(D-lactide-co-glycolide) to reduce osmotic damage to encapsulated cells.

One of the major shortcomings of this approach is the controlled positioning of multiple cell types in a particular location, as the mixture of hydrogel precursor and tissue-specific cells is poured manually to encapsulate the 3D vascular network. The removal of sacrificial filaments from scaffolds is also challenging as it might require cytotoxic solvents, high temperature, or elevated pressure. Moreover, cell viability in the scaffold during UV crosslinking could be significantly compromised due to the cytotoxic effect of photo radiation or photo initiator.

**Example 7.2**

A vascular network is to be created within a scaffold by extruding sodium alginate (3% w/v) as sacrificial strands by means of a 3D bioplotter.

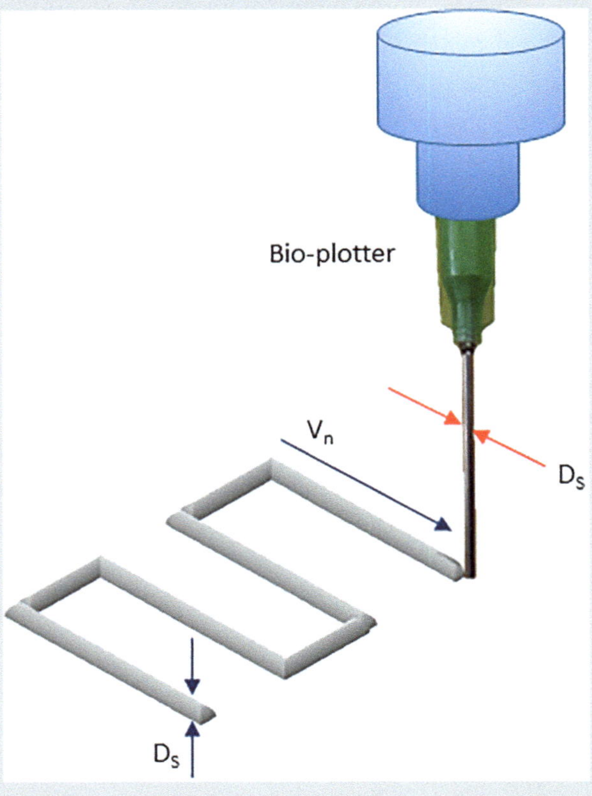


(continued)

**Example 7.2** (continued)

The mass flow rate of the hydrogel precursor is 1.25 mg/s, the linear needle speed is 14 mm/s, and the equation for calculating strand diameter, as introduced in Chap. 5, is

$$Ds = \sqrt{\frac{4Q_h}{\pi V_n}},$$

where $D_s$, $Q_h$, and $V_n$ are the strand diameter, hydrogel precursor flow rate, and needle speed, respectively. Calculate the diameter of the extruded strand.

**Solution**

Weight of 100 mL 3% (w/v) sodium alginate solution is $(100 + 3)$ g $= 103$ g.

Density of 3% (w/v) sodium alginate solution is $(103$ g/100 mL$) = 1.03$ g/mL $= 1.03$ mg/mm$^3$.

Sodium alginate flow rate $(1.25$ mg/s $\div 1.03$ mg/mm$^3) = 1.21$ mm$^3$/s.

Therefore, strand diameter, $D_s = \sqrt{\frac{4Q_h}{\pi V_n}} = \sqrt{\frac{(4)(1.21 \text{ mm}^3/\text{s})}{(3.14)(14 \text{ mm}/\text{s})}} = 0.506$ mm **(Ans.)**

**Example 7.3**

You need to fabricate an $8 \times 8 \times 6$ mm$^3$ vascular network with sacrificial strands as per the following specifications:

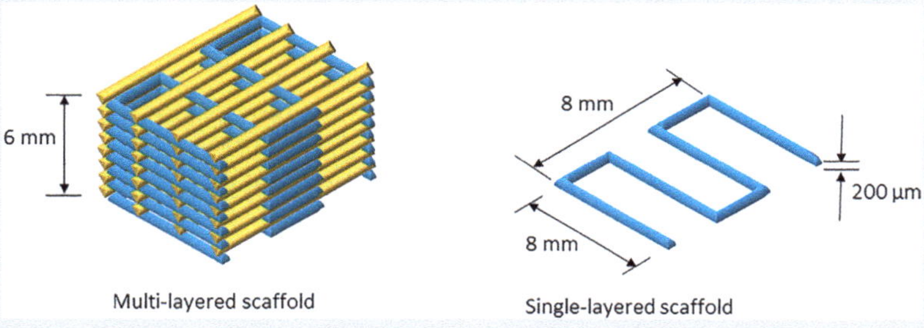


(a) Diameter of strands $= 200$µm.
(b) Number of strands in each layer $= 4$.
(c) Density of sacrificial hydrogel precursor $= 1.4$ g/mL solution of sacrificial polymer (solvent was water).
(d) Dissolution capacity of solvent $= 0.25$ mg sacrificial polymer/mL solvent.

Estimate the required amount of sacrificial polymer and solvent to dissolve the sacrificial network assuming that the extruded strands are cylindrical.

**Solution**

Total number of 2D layers $= (6 \text{ mm} \div 200\mu\text{m}) = (6 \text{ mm} \div 0.2 \text{ mm}) = 30$.

Number of sacrificial strands in each layer $= 4$.

Total length of sacrificial strands in each layer $= (4 \times 8) \text{ mm} = 32 \text{ mm}$.

Total length of sacrificial strands in whole scaffold $= (32 \times 30) = 960 \text{ mm}$.

Required volume of sacrificial polymer $= \pi \text{ (strand radius)}^2 \text{ (strand length)} = (3.14) (0.1 \text{ mm})^2 (960 \text{ mm}) = 30.14 \text{ mm}^3 = 0.0301 \text{ mL}$.

Required mass of sacrificial polymer $= (0.030 \text{ L ml solution of sacrificial polymer}) (0.4 \text{ g sacrificial polymer/1 mL solution of sacrificial polymer}) = 0.012 \text{ g}$ **(Ans.)**

Required volume of dissolution solvent $= (12 \text{ mg of sacrificial polymer}) \div (0.25 \text{ mg sacrificial polymer/mL dissolution solvent}) = 48 \text{ mL}$ **(Ans.)**

### *7.4.3  Self-Assembled Vasculature Using Bioprinting*

A wide range of natural, synthetic, and hybrid biomaterials have been used in bioprinting; however, none are free from shortcomings. Most biomaterials show uncontrolled degradation, immunogenicity, inflammation, and cytotoxicity during *in vivo* or in vitro applications. In some cases, biomaterials inhibit ECM secretion, distribution, and organization as well as cell-cell communication. To eliminate the complexities of biopolymers, a self-assembly approach has been developed as an alternative, where scaffold-free multicellular spheroids or filaments are extruded using a bioprinter to form tissue engineered scaffolds. In such systems, spheroids or sacrificial filaments are printed layer-by-layer concurrently with various cell-incorporated strands as designed.

In the preparation of self-assembled tissue strands, a hydrogel-based luminal tube is fabricated using a coaxial extrusion system to store and culture cell aggregations for a specific period. The diameter of the luminal tube depends on the choice of fabrication parameters (e.g., extrusion pressure, temperature, needle opening). Generally, a microsyringe is used to load EC or EC/mixed tissue-specific cell pellets in the luminal tube. Upon cell loading, vascular clamps are used to tie both ends of the tubular conduit and the cell pellets are cultured for a couple of days. Being semipermeable, the luminal tube does not allow cells to leave the conduit but facilitates nutrient transfer from the surrounding culture media to the loaded cell population. Notably, the hydrogel precursor selected in the tube fabrication is inert to cell attachments, thus compelling the loaded cells to form vascularized tissue over time. In this approach, cell viability might be compromised during microinjection and the tissue formed in the luminal tube might contract in the radial direction. When tissue self-assembles inside the luminal tube, an appropriate solution (e.g., sodium citrate, EDTA) is used to dissolve the hydrogel conduit and release the vascularized tissue strand [19]. Eventually, a customized multi-arm bioprinter is used to fabricate scaffolds using the self-assembled tissue strands that mimic native tissue (Fig. 7.7a).

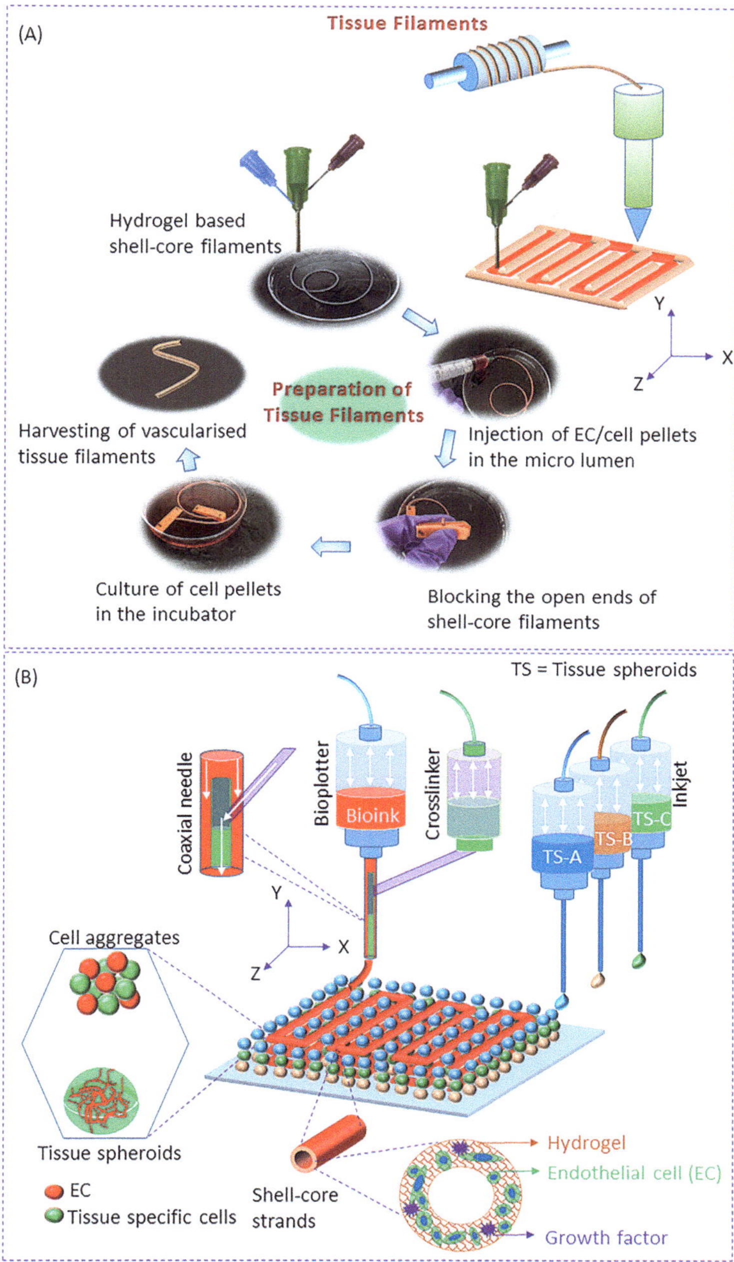


**Fig. 7.7** Formation of vascular network with self-assembled structure: (**a**) tissue filaments, (**b**) tissue spheroids or cell aggregates

To form vascularized tissue spheroids, cell pellets are first prepared by centrifuging an EC and tissue-specific cell mixture. The cell pellets are then transferred into capillary micropipettes and incubated for a given period during which they form sausage-like structures. These sausage-like structures are extruded from the micropipette as filaments and sliced into multiple pieces of similar diameter and length that eventually form spheres. The multicellular spheroids prepared in this way are loaded in cartridges for biofabrication. Using an extrusion-based (EB) system, shell-core filaments and vascularized tissue spheroids are fabricated side by side to facilitate formation of an interconnected and perfused vascular network (Fig. 7.7b).

Scaffold-free approaches to fabricate macro blood vessels have received significant attention in recent years. In these approaches, sacrificial strands are printed inside and outside the blood vessel to provide channels and structural support, respectively [20]. Typically, blood vessels are fabricated with tissue spheroids or cell aggregates. Briefly, to prepare a cylindrical bioink, EC- and SMC-based cylindrical filaments are extruded into non-adhesive agarose or Teflon™ molds using a bioprinter and incubated overnight to add cohesive properties to the cellular cylinders. To prepare sacrificial bioink, micropipettes filled with a sacrificial polymer (e.g., liquid agarose) are dipped into cold phosphate-buffered saline (PBS) to facilitate hydrogel formation. The agarose gels at the cold temperature and demonstrates outstanding extrusion properties, being non-adhesive to the micropipette [21]. The tissue and sacrificial filaments are loaded into a customized bioprinter that fabricates the agarose and tissue filaments in a layer-by-layer fashion to generate single-layered (i.e., EC filaments) or double-layered (i.e., EC and SMC filaments) vascular constructs (Fig. 7.8). After incubation in a bioreactor, the sacrificial strands disappear while multicellular spheroids or tissue filaments fuse together to form single- or double-layered micro- or macrovascular tubes within a week.

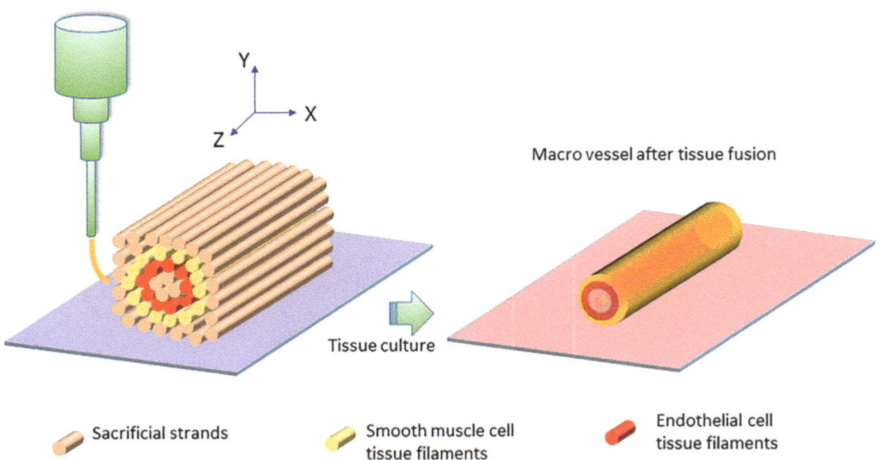


**Fig. 7.8** Fabrication of macro blood vessel using sacrificial filaments

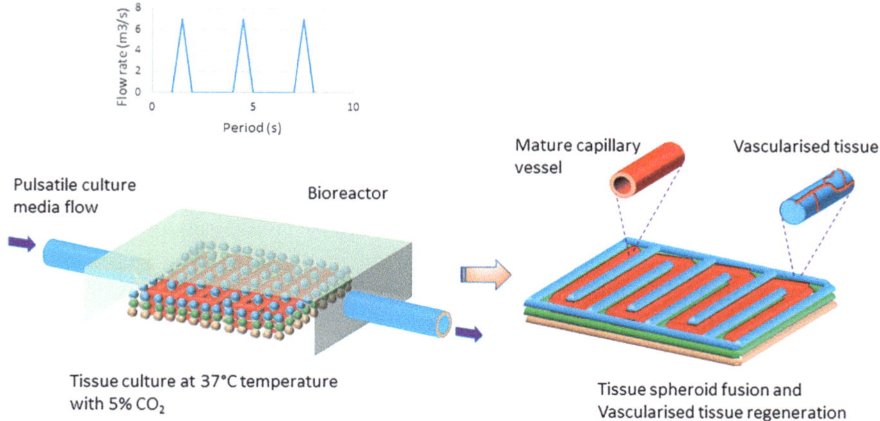



**Fig. 7.9** Fusion of tissue spheroids in a bioreactor under pulsatile flow conditions

*In vitro*, a bioreactor can be used to maintain pulsatile media flow through the vascular network to facilitate the formation of a lumen-like structure. Tissue culture media (i.e., DMEM) are often used to flow through the vascular network to supply nutrients, growth factors, and oxygen gas to the large incorporated cell population. The efficiency of diffusion mass transfer depends on the flow velocity and concentration of media as well as the relative diffusivity of various biomolecules. Because fluid velocity is associated with the wall shear stress, the fluid flow rate should be chosen carefully to find the optimum mass transfer rate and shear stress required to grow a functional vasculature network (Fig. 7.9).

## 7.5   Other Vascularization Approaches

Apart from the EB approach, laser-based approaches are an alternative way of fabricating vascularized 2D/3D tissue constructs in a layer-by-layer fashion. Laser-induced forward transfer (LIFT), matrix-assisted pulsed laser evaporation direct writing (MAPLE DW), and stereolithography are well-known techniques for printing 2D/3D cell patterns along with a capillary network. In the LIFT/MAPLE DW technique, projected laser pulses on the donor material create bubbles that advance beneath the cell/polymer mixture and deposit a micro-pattern on the receiver substrate. Application of the LIFT and MAPLE DW technique eliminates the problem of fabrication nozzle clogging and high-viscosity bioink handling, while ensuring high-resolution printing with outstanding accuracy. Laser-based stereolithography (LS) fabricates large scaffolds with detailed vascular networks in a layer-by-layer fashion in the presence of photosensitive materials and photo-initiator. Although laser-based approaches are useful in rapid prototyping any 3D structure with precision, laser-induced cell damage and cytotoxicity of the photo-initiator or resin need to be taken into consideration before future applications.

Alternative approaches to 3D fabrication have evolved to achieve better control of spatial geometry. Stacking multiple micro-patterned 2D planar substrate results in the formation of an intricate 3D vascular network. Replica molding, soft lithography, plasma etching, laser ablation, and direct laser lithography are the techniques that are useful in the creation of micro-patterned 2D substrates. After fabrication, ECs are seeded in the patterned substrate using a microfluidic approach and cultured in a bioreactor while maintaining appropriate shear stress. Biodegradable polymers (e.g., silk fibroin, Matrigel™, type I collagen, fibrin) demonstrate superior biocompatibility and mass diffusivity compared to synthetic polymers (e.g., polystyrene, polycarbonate, silicon, polyvinyl chloride) for the generation of micro-patterned substrates. Notably, preparation of vascularized 3D scaffolds by assembling micro-patterned 2D planar surfaces is a time-consuming process and associated with stacking complexities.

Assembly of cell-loaded micromodules in a closed vessel results in the formation of vascularized tissue upon culture in the bioreactor. The micromodules, prepared with a microscale molding technique, often accumulate following random distribution, gravity-enforced self-assembly, or directed self-assembly approaches. To mimic native tissue, tissue-specific cells are incorporated in the micromodules, while ECs are used to coat the outer surface. Such micromodules grow capillary blood vessels in the interstitial space in a random fashion upon perfusion with blood or culture medium. Furthermore, shape-controlled microgels are useful in the formation of vasculature having linear, branched, or offset geometry. However, macroscale vascularized tissue grown using this approach demonstrates poor tissue integration and vasculature formation in vitro and *in vivo*.

*In vivo*, ECs interact with the surrounding ECM during the development of vascular networks. Electrospinning techniques facilitate the fabrication of nano fibers mimicking the structure of natural ECM. In particular, a charged nano-scale polymer jet is generated by applying a large voltage difference between the polymer solution and a fiber collector, and thus the charged jet accumulates on the collector as a nanofiber. Incorporation of nanofibers in the scaffolds promotes vascularization interacting with ECs both in vitro and *in vivo*. Many synthetic (e.g., PLLA, PCL, PVA, PLGA, PEO) and natural (e.g., collagen, chitosan, gelatin) polymers allow for nano-fabrication with the electrospinning technique. Synthetic polymers have better mechanical stability and fabrication flexibility compared to natural polymers; however, natural polymers are preferable over synthetic polymers in terms of biological performance and biodegradability. In contrast, nanofibers prepared from copolymers, composite, or hybrid polymers demonstrate the desired mechanical and biological performance simultaneously.

*In vivo*, the interactions between cells and ECM play a vital role in the tissue regeneration process. Mimicking the biochemical, biophysical, and topographical cues embedded in natural ECM remains a challenge in tissue engineering. In contrast, decellularized tissues have the ability to provide all of the cues necessary for capillary network formation in a superior fashion to biofabricated scaffolds.

Several physical (e.g., electroporation, pressure, temperature), chemical (e.g., detergent, base, acid), and biological (e.g., dispase and trypsin) methods can be used for tissue decellularization. The selection of a decellularizing agent depends on the size, density, thickness, and lipid content of the specific tissue. To vascularize a decellularized matrix, ECs are seeded through a specific artery or vein either manually or with the help of a bioreactor. The reseeded ECs eventually form a vascular network through attachment, proliferation, and differentiation. Although promising, tissue decellularizing processes are not free from shortcomings. Some common outcomes of the tissue decellularization process include destruction of the ECM ultrastructure, removal of ECM protein and growth factors, and immunological complexities; these need to be addressed for possible future applications.

## 7.6 Summary

*In vivo* vascular networks are composed of arteries and veins. Arteries branch out into arterioles, metarterioles, and capillaries to supply nutrients and oxygen to surrounding cells. The capillaries combine downstream to form veins. Angiogenesis and vasculogenesis are the two mechanisms by which blood vessels form *in vivo*. In angiogenesis, new blood vessels grow from existing ones in response to VEGF and invade the ischemic tissue. In vasculogenesis, EPCs accumulate at the ischemic site in response to cytokines and growth factors and eventually a blood vessel forms.

Shell-core strands printed from co-axial needles demonstrate a lumen-like structure and form 3D vascular networks upon culture in a bioreactor. The addition of growth factors, peptides, proteins, or genetically modified cells further enhances the biological performance of shell-core strands. Incorporation of 3D sacrificial networks of strands in the scaffolds are sometimes effective to overcome the difficulties in directly bioprinted vascular networks. Upon encapsulating the network with hydrogel(s), the sacrificial strands are removed using an appropriate solvent or temperature, and then the remaining hollow lumens are seeded with ECs within a bioreactor.

Tissue spheroids or filaments are a better choice to promote vascularization compared to biopolymer-included strands. Culturing tissue-specific cells along with ECs results in vascularized cell aggregates or tissue filaments. The self-assembled structure further anastomoses with shell-core strands and forms a perfused capillary network.

In addition to extrusion bioprinting approaches, vascular networks can be fabricated using laser and electrospinning techniques. Stacked multiple micro-pattern substrates or micro-modules have the ability to facilitate the growth of vasculature. Decellularized natural tissue or organs have the potential to grow tree-like vasculature without immunological rejections.

**Problems**

1. Briefly explain why vascular networks are needed within large and thick tissue scaffolds.
2. Briefly explain the composition and function of vascular networks *in vivo*.
3. Explain the *in vivo* blood vessel formation mechanisms of angiogenesis and vasculogenesis.
4. What are the common components of bioinks used for printing vascular scaffolds? From the literature, name one bioink for this purpose and briefly explain each component used in the bioink.
5. How are shell-core strands used to form a vasculature network?
6. Briefly describe vascular network formation by a 3D sacrificial network.
7. How are vascular networks created using self-assembled strands or tissue spheroids?
8. You need to fabricate a $10 \times 10 \times 5 \ mm^3$ vascular network with a coaxial needle as per the following specifications:

   (a) Radius of shell $= 200 \mu m$.
   (b) Radius of core $= 100 \mu m$.
   (c) Number of strands in each layer $= 10$.
   (d) 3% (w/v) hydrogel precursor solution (solvent was water)
   (e) Required growth factor $= 10$ mg growth factor/mL hydrogel precursor solution.

   Estimate the required amount of biopolymer and growth factor to fabricate the scaffold.

# References


1. H. Sarin, Physiologic upper limits of pore size of different blood capillary types and another perspective on the dual pore theory of microvascular permeability. J. Angiogenes. Res. BioMed Central **2**, 14 (2010)
2. P.C. Maisonpierre, C. Suri, P.F. Jones, S. Bartunkova, S.J. Wiegand, C. Radziejewski, et al., Angiopoietin-2, a natural antagonist for Tie2 that disrupts *in vivo* angiogenesis. Science **277**, 55–60 (1997)
3. I. Kim, H.G. Kim, S.-O. Moon, S.W. Chae, J.-N. So, K.N. Koh, et al., Angiopoietin-1 induces endothelial cell sprouting through the activation of focal adhesion kinase and plasmin secretion. Circ. Res. Am Heart Assoc **86**, 952–959 (2000)
4. A.R. Nelson, B. Fingleton, M.L. Rothenberg, L.M. Matrisian, Matrix metalloproteinases: Biologic activity and clinical implications. J. Clin. Oncol **18**, 1135 (2000)
5. J.A. Belperio, M.P. Keane, D.A. Arenberg, C.L. Addison, J.E. Ehlert, M.D. Burdick, et al., CXC chemokines in angiogenesis. J. Leukoc. Biol. Soc Leukocyte Biol. **68**, 1–8 (2000)
6. P. Carmeliet, R.K. Jain, Angiogenesis in cancer and other diseases. Nature **407**, 249 (2000)
7. C. Suri, J. McClain, G. Thurston, D.M. McDonald, H. Zhou, E.H. Oldmixon, et al., Increased vascularization in mice overexpressing angiopoietin-1. Science **282**, 468–471 (1998)
8. C. Urbich, S. Dimmeler, Endothelial progenitor cells. Circ. Res. Am. Heart Assoc. **95**, 343–353 (2004)

9. H. Spring, T. Schüler, B. Arnold, G.J. Hämmerling, R. Ganss, Chemokines direct endothelial progenitors into tumor neovessels. Proc. Natl. Acad. Sci. USA **102**, 18111–18116 (2005)
10. P.-H. Huang, Y.-H. Chen, C.-H. Wang, J.-S. Chen, H.-Y. Tsai, F.-Y. Lin, et al., Matrix metalloproteinase-9 is essential for ischemia-induced neovascularization by modulating bone marrow–derived endothelial progenitor cells. Arterioscler. Thromb. Vasc. Biol. Am Heart. Assoc. **29**, 1179–1184 (2009)
11. P. Hildbrand, V. Cirulli, R.C. Prinsen, K.A. Smith, B.E. Torbett, D.R. Salomon, et al., The role of angiopoietins in the development of endothelial cells from cord blood CD34+ progenitors. Blood. Am. Soc. Hematol. **104**, 2010–2019 (2004)
12. W. Suh, K.L. Kim, J. Kim, I. Shin, Y. Lee, J. Lee, et al., Transplantation of endothelial progenitor cells accelerates dermal wound healing with increased recruitment of monocytes/ macrophages and neovascularization. Stem Cells **23**, 1571–1578 (2005)
13. M.D. Sarker, S. Naghieh, N.K. Sharma, L.Q. Ning, X.B. Chen, Bioprinting of vascularized tissue scaffolds: Influence of biopolymer, cells, growth factors, and gene delivery. J. Healthc. Eng **2019**, 1 (2019)
14. M. Izadifar, M. Kelly, X.B. Chen, Regulation of sequential release of growth factors using bi-layer polymeric nanoparticles for cardiac tissue engineering. Nanomedicine **11**(24), 3237–3259 (2016)
15. M.D. Sarker, M. Izadifar, D.J. Schreyer, X.B. Chen, Influence of ionic crosslinkers ($Ca^{2+}$/ $Ba^{2+}$/ $Zn^{2+}$) on the mechanical and biological properties of 3D boplotted hydrogel scaffolds. J. Biomater. Sci. Polym. Ed. **29**, 1126–1154 (2018)
16. W. Liu, Z. Zhong, N. Hu, Y. Zhou, L. Maggio, A.K. Miri, et al., Coaxial extrusion bioprinting of 3D microfibrous constructs with cell-favorable gelatin methacryloyl microenvironments. Biofabrication **10**, 24102 (2018)
17. D. Therriault, R.F. Shepherd, S.R. White, J.A. Lewis, Fugitive inks for direct-write assembly of three-dimensional microvascular networks. Adv. Mater. **17**, 395–399 (2005)
18. D.B. Kolesky, R.L. Truby, A. Gladman, T.A. Busbee, K.A. Homan, J.A. Lewis, 3D bioprinting of vascularized, heterogeneous cell-laden tissue constructs. Adv. Mater. **26**, 3124–3130 (2014)
19. Y. Yu, K.K. Moncal, J. Li, W. Peng, I. Rivero, J.A. Martin, et al., Three-dimensional bioprinting using self-assembling scalable scaffold-free "tissue strands" as a new bioink. Sci. Rep. **6**, 28714 (2016)
20. M.D. Sarker, S. Naghieh, N.K. Sharma, X.B. Chen, 3D biofabrication of vascular networks for tissue regeneration: A report on recent advances. J. Pharm. Anal. **8**, 277–296 (2018)
21. C. Norotte, F.S. Marga, L.E. Niklason, G. Forgacs, Scaffold-free vascular tissue engineering using bioprinting. Biomaterials **30**, 5910–5917 (2009)

# Chapter 8
# Controlled Release of Biomolecules in Printed Scaffolds

## 8.1  Introduction

In tissue engineering, cells from patients or other sources are incorporated into engineered tissue scaffolds with the addition of bioactive molecules (or biomolecules), such as growth factors (GFs), to trigger/promote cellular growth and/or functions. The tissue scaffolds, once implanted into the damaged tissue/organ site, aid in cellular regeneration and recovery of tissue/organ function, with cell attachment, proliferation, and differentiation being key to successful tissue regeneration. GFs and other biomolecules activate sequential intracellular signaling pathways and control cellular gene expression. As such, the dose, type, release rate, and timing of biomolecule availability affect the cellular response and function and thus must be appropriately regulated to promote regeneration of the desired tissue. This chapter introduces the concept of controlled release and the common strategies for regulating biomolecules in tissue engineering, along with the common biomolecules and their controlled release modes. It also discusses various methods to incorporate or load biomolecules into engineered tissue scaffolds and to achieve controlled release via micro/nanoparticles.

## 8.2  Controlled Release of Biomolecules

In tissue engineering, controlled release of biomolecules refers to the provision of biomolecules in a controllable manner. Figure 8.1 shows the release profiles of biomolecules for cases both with and without controlled release, indicating the number of biomolecules released verses time and demonstrating how many of the biomolecules come into action. The therapeutic window refers to the concentration range of biomolecules that is appropriate for cell function and tissue regeneration at


D. X. B. Chen, *Extrusion Bioprinting of Scaffolds for Tissue Engineering*,
https://doi.org/10.1007/978-3-031-72471-8_8

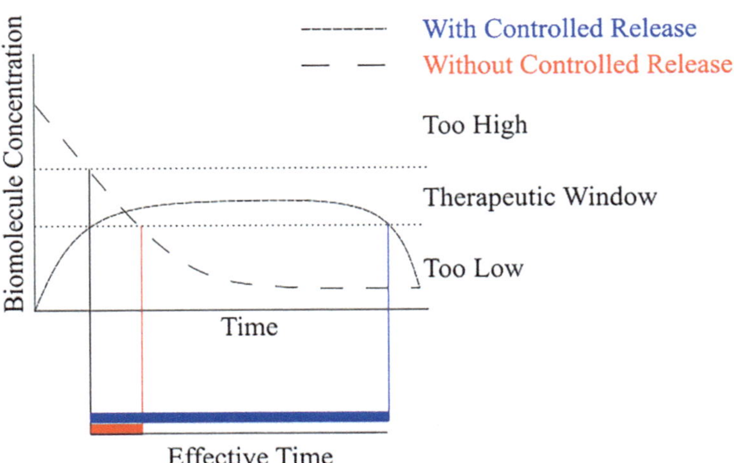


**Fig. 8.1** Release profiles of biomolecules with and without controlled release

the site of damage or injury, i.e., neither so little as to be ineffective nor so much as to result in toxic side effects.

First, let's examine the profile without controlled release. At the onset, many biomolecules are released and are available at a concentration greater than the upper limit of the therapeutic window, which can cause toxic side effects. The release profile then quickly passes through the therapeutic window and enters the zone of ineffective concentration, i.e., where bioactivity is too low. This is analogous to the situation in which patients take their oral medicine or pharmaceutical; the drug quickly disperses in their body and passes through the therapeutic window, and they are then required to take another pill after a certain time period. In contrast, the profile with controlled release shows the concentration of the biomolecule swiftly rises to the lower limit of the therapeutic window and remains within the effective range for a longer time period compared to the scenario without controlled release. The goal of controlled release in tissue engineering is to mimic the biological release profiles of biomolecules so as to promote cell functions and tissue regeneration. Designing controlled release systems that properly incorporate biomolecules can achieve gradual release of the bioactive molecules from the scaffold at a concentration within the therapeutic window over a prolonged period of time, thus prolonging or eliciting the desired cell functions and tissue regeneration.

The design of a controlled release system is of great importance for regulating the induction of biological activities, including cellular adhesion, proliferation, differentiation, and migration. This controlled biological activity is beneficial in vitro, and is usually essential in vivo. For in vitro studies on scaffolds with cells, the desired biological activities might be achieved by adding biomolecules or signals to the cell culture medium in the bioreactor, and, in such cases, controlled release may not be required. In some cases, controlled release may be desired to provide biomolecules or signals locally within the scaffold. For in vivo studies, controlled release is essential due to the limited ability to intervene/interfere once the scaffold is implanted. Stimulation of cellular activities can be achieved through the addition and release of GFs such as bone morphogenetic proteins (BMPs), transforming growth factors (TGFs), fibroblast growth factors (FGFs), vascular endothelial growth factors (VEGFs), and adhesion factors, such as fibronectin, vitronectin, and laminin. These GFs are large, water-soluble molecules that are highly sensitive; minute alterations to the environment can cause their denaturation or inactivity [1, 2]. Signaling factors should be provided in their active and appropriate state; for instance, in a physiological system, adhesion factors are immobilized and present in the extracellular matrix (ECM) for a desired period of time, while GFs are active in a diffusible state that is then bound and internalized by cells. To this end, systems designed to achieve in vivo controlled release must be evaluated and validated in vitro to ensure the biological release profiles and bioactivity functions as desired for promoting cell functions and tissue regeneration.

## 8.3  Biomolecules and Controlled Release Modes

### 8.3.1  Biomolecules in Tissue Engineering

The term biomolecule, as used in this chapter, refers to small bioactive molecules (often proteins) such as GFs, adhesion factors, cytokines, and other molecules that stimulate cell functions, tissue regeneration, and/or immune activities. All of these molecules have varying physiological roles in promoting cellular activities.

GFs are molecules capable of stimulating complex signal cascades that guide and promote cellular growth and differentiation of specific cell types. In general, GFs are ligands for cell-surface receptors that interact with receptor tyrosine kinases [1]. The interaction of GFs induces dimerization and phosphorylation of the receptor, which leads to further alteration of intracellular proteins that subsequently trigger a signaling cascade. Families of GFs commonly used in tissue engineering include VEGFs, BMPs, TGFs, and FGFs, among others, with some examples shown in Table 8.1. The effect of GFs on a physiological system is dependent on the dose, type, release rate, and timing of GF release, with interplay between the release of different GFs leading to the generation of tissues. For example, angiogenesis, or the formation of new blood vessels as discussed in Chap. 7, is initiated and moderated by VEGF and basic FGF, while platelet-derived growth factor (PDGF) controls the maturation of newly formed blood vessels [2]. GFs exhibit a short physiological half-life (~ minutes), and degrade rapidly in cellular microenvironments, making their controlled release and the maintenance of a specific level of bioactivity important for maintaining cellular growth and differentiation [3, 4].

Adhesion factors are natural components of the ECM and function as an interface between cellular materials and the ECM. These factors are present in high concentrations in proteins such as fibronectin, fibrinogen, and laminins, among many others, and are functional in an immobile state [1]. Smaller integrin-binding domains

**Table 8.1** Examples of commonly used growth factors in tissue engineering

| Tissue engineering application | Growth factors |
| --- | --- |
| Angiogenesis | VEGF, FGF, PDGF, bFGF, Ang |
| Bone | TGF-β, PDGF, rhBMP, rhIGF |
| Cartilage | TGF-β, BMP |
| Dermis | EGF, bFGF |

Note: *VEGF* vascular endothelial growth factor, *FGF* fibroblast growth factor, *PDGF* platelet-derived growth factor, *bFGF* basic fibroblast growth factor, *Ang* angiopoietin, *TGF-β* transforming growth factor-β, *rhBMP* recombinant human bone morphogenetic protein, *rhIGF* recombinant human insulin-like growth factor, *BMP* bone morphogenetic protein, *EGF* epidermal growth factor

can be isolated from these whole proteins and incorporated into other materials such as tissue scaffolds to promote cellular adhesion and migration within the scaffolds. Adhesion factors bind to transmembrane receptors, allowing for interaction of the intracellular domain and cytoskeleton through intermediary proteins [1].

Cytokines are small proteins that affect cellular behaviors, often in roles related to innate and acquired immunity. Classes of cytokines include chemokines, interferons, interleukins, and tumor necrosis factors (TNFs). Chemokines are small signaling proteins secreted by cells that are vital for directing cell migration. A chemokine gradient promotes chemotaxis, or movement of specific cell types up a chemokine concentration gradient, in the case of injury or infection, to ensure the required cell types are present at the affected site. Interferons are signaling proteins induced by virus-infected monocytes and lymphocytes that interfere with viral replication and warn neighboring cells of the presence of a virus. Interleukins promote the proliferation of various types of immune cells and play a role in the inflammatory response, antibody production, and further activation of immune cells. TNFs also play a role in the inflammatory response, often promoting fever, stimulation of further production of cytokines, and activation of fibroblasts. Various other bioactive molecules have been investigated for controlled release from tissue scaffolds, including immunological molecules such as immunoglobulins (IgG, IgA, etc.) and small therapeutic drugs [5, 6].

### 8.3.2 Modes of Controlled Release

Models of controlled release are associated with the ways to incorporate biomolecules into biomaterials. Biomolecules are generally incorporated in the homogeneous or heterogeneous way. In the homogeneous way, the loaded biomolecule is soluble in the selected biomaterial; in the heterogeneous way, water-soluble biomolecules are dispersed in hydrophobic polymers in which they are not soluble. Contingent upon interactions among the bioactive molecules and biomaterial as well as the mechanism of material degradation, more than one mode of biomolecule release usually exists. Commonly, models of controlled release are classified as

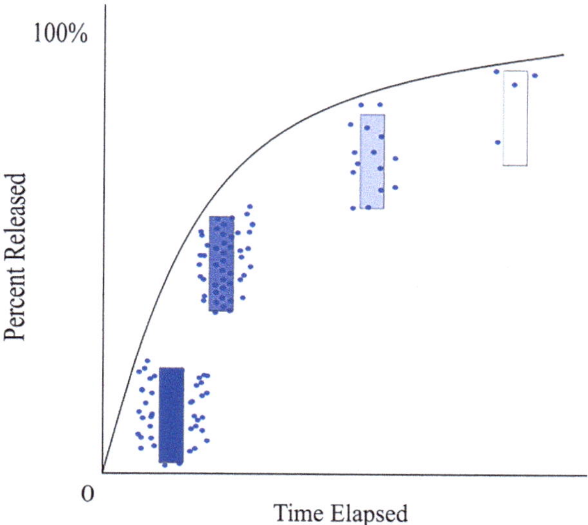



**Fig. 8.2** Diffusion-controlled release and its associated release profile

diffusion-controlled release, homogeneous degradation, surface erosion, and biomolecule immobilization.

Diffusion-controlled release occurs when the matrix material has a molecular-chain mesh size larger than that required by the biomolecule for diffusion. In these cases, the loaded biomolecule is capable of diffusing freely out of the matrix material into the environment, and release occurs independent of matrix material degradation. The release rate in diffusion-controlled systems is not constant but depends on time. As shown in Fig. 8.2, the diffusion rate is initially at its maximum value and then declines over time as the biomolecule concentrations internal and external to the scaffold become more similar. This mode may be useful when fast initial release is desired; however, it lacks the ability to sustain release at a consistent rate or within a desired therapeutic window.

If the mesh size of the matrix material is too small to allow for efficient diffusion of the loaded biomolecule, the release rate is dependent on the degradation of the biomaterial. Homogeneous degradation release is caused by hydrolysis occurring throughout the bulk of the matrix material due to environmental interactions, as shown in Fig. 8.3. In concert with swelling of the material, homogeneous degradation leads to the widening of pores throughout the entirety of the construct, allowing for the release of the incorporated biomolecules. Due to the reliance on bulk degradation, initial release is limited before increasing once sufficient degradation has occurred. At this stage, the release rate is controlled by a balance of diffusion and further bulk degradation. As increased degradation frees more of the loaded bioactive molecule, the diffusion rate decreases as the concentration of the biomolecule within the scaffold decreases. If properly balanced, controlled release through homogeneous degradation can be used to achieve zero-order kinetics, where the release rate is constant over an extended period of time. This form of release is

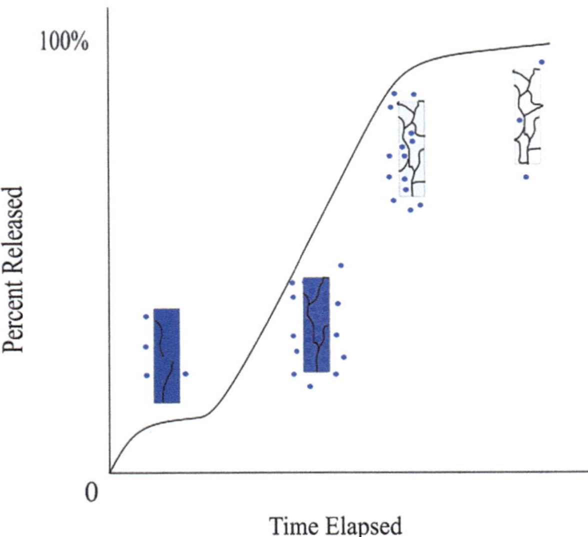

**Fig. 8.3** Homogeneous or bulk degradation release and its associated release profile

known to occur in materials such as polylactic acid (PCL), polyglycolic acid (PGA), and poly(lactide-co-glycolide) acid (PLGA).

For more hydrophobic materials, environmental interaction tends to be limited to the external surface of the engineered construct. As surface erosion occurs, biomolecule release occurs from the degraded surface, as shown in Fig. 8.4. Control over the release rate can be exercised through the selection of the matrix material for a specific degradation rate and tailoring of the construct design to one that maintains a consistent surface area as surface degradation occurs, allowing for attainment of zero-order release kinetics.

Biomolecule immobilization is used less frequently than the other modes of controlled release and is employed when no release is desired. Although this may seem counterintuitive, biomolecule immobilization allows for the extended presentation of molecules such as adhesion factors that are functional in the immobile state. As surface and bulk degradation of the matrix material will still occur over time, matrix materials are often selected to have a degradation rate similar to the time for which immobilized bioactive molecules are desired to be present on the construct. The effect of degradation of the surface on which the biomolecules are immobilized can be minimized through incorporation of the bioactive molecule throughout the matrix, leading to further exposure as the surface degrades. As such, bulk degradation will limit the duration of immobilization of adhesion molecules.

While these modes of release have been discussed individually, in most scenarios a combination of release mechanisms plays a role in the overall release kinetics of biomolecules. Determining and tailoring the dominance of these release modes allows for optimal release kinetics to be realized for different applications. Other considerations related to release mechanisms include ensuring the incorporated

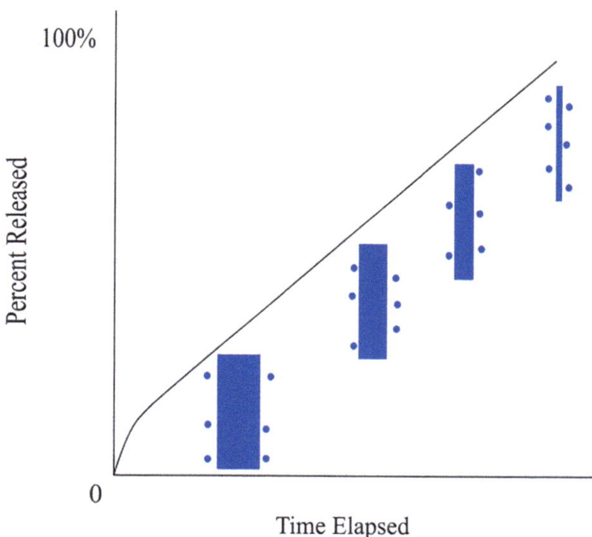


**Fig. 8.4** Surface degradation release and its associated release profile

biomolecules do not stimulate an unwanted immune response, as they may only be partially exposed and, thus, unrecognizable to the host immune system at certain points of release [1]. As the generation of an immune response to naturally occurring biological molecules present in the host could cause severe auto-immune issues, these concerns require careful attention when designing a controlled release system.

## 8.4   Loading of Biomolecules into Scaffolds

Along with the different modes of biomolecule release from engineered constructs are varying methods or systems to incorporate or load the selected biomolecules into scaffolds. The selection of loading method is related to interactions between the biomolecules and the selected encapsulating material, with common methods classified as chemical immobilization, physical encapsulation, or loading in micro/nanoparticulate systems.

### *8.4.1   Chemical Immobilization*

Figure 8.5 is a schematic diagram of chemical immobilization of GFs/cytokines into an engineered or printed scaffold. By this method, GFs or other bioactive molecules are chemically conjugated within the scaffold biomaterial. The type of chemical binding (non-covalent or covalent) and affinity interaction between the biomolecules

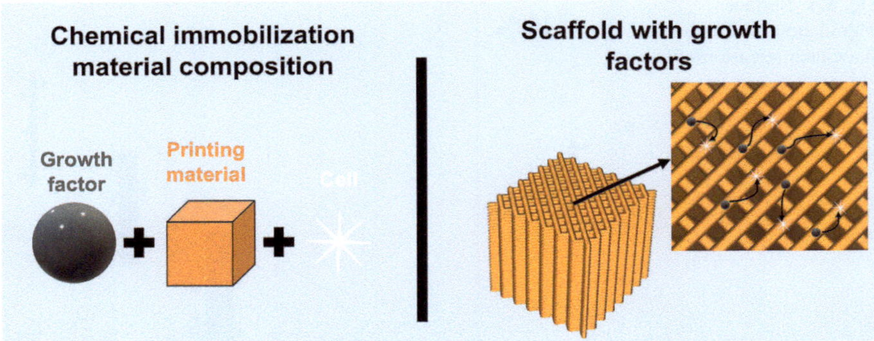


**Fig. 8.5** Schematic diagram of chemical immobilization of growth factors within a scaffold

and the scaffold have an impact on the release rate of the biomolecules. In non-covalent conjugation, the biomolecules are incorporated by electrostatic (charge–charge) interaction or indirect interactions via intermediate proteins such as fibrin, chondroitin, and glycosaminoglycans, among others. In covalent conjugation, the bioactive molecules are conjugated by a functional group that provides a more prolonged release. However, the functional sites of biomolecules such as GFs can be damaged during chemical immobilization, leading to reduced bioactivity. In addition, chemically immobilized bioactive molecules are dispersed throughout the scaffold and do not provide spatial concentration gradients to achieve cellular migration and proliferation.

## 8.4.2 Physical Encapsulation

Figure 8.6 is a schematic diagram of the physical encapsulation of GFs into a scaffold. Physical encapsulation of biomolecules in scaffold biomaterials is a more flexible mechanism for controlling the release rate as it relies on regulating the molecular weight, hydrophobicity, and porosity of the biopolymer. The presentation pattern of the incorporated molecules is regulated by the degradation rate of the scaffold matrix, diffusivity of the biomolecules, and composition/structure of the scaffold. GFs and other biomolecules can be physically incorporated into the scaffold by immersing the pre-formed scaffolds in a solution containing the biomolecules to be loaded, allowing for their diffusion into or entrapment in the scaffold. Diffusion of the biomolecules through the scaffold pores creates concentration gradients throughout the scaffold that help regulate cell migration. Drawbacks of this method include low loading capacities, long loading times, and burst release profiles. In addition, exposure of the loaded GFs to the microenvironment may result in their cleavage or destruction by hydrolytic enzymes or oxidation.

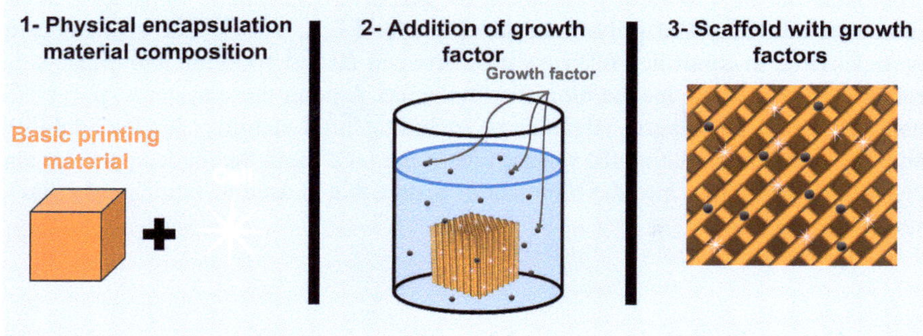



**Fig. 8.6**  Schematic diagram of physical encapsulation of growth factors within a scaffold

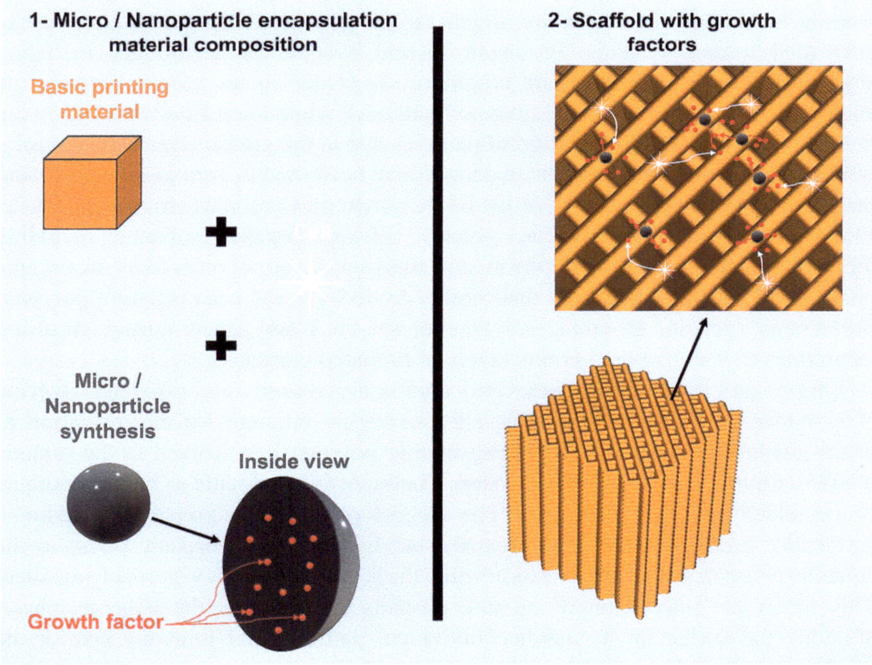



**Fig. 8.7**  Schematic diagram of the encapsulation of bioactive molecules in particles and subsequent incorporation into a scaffold

### 8.4.3   Micro/Nanoparticles

Another controlled release approach is to employ micro/nanoparticle delivery systems that protect the incorporated GFs and modulate the release profiles. Figure 8.7 is a schematic of the incorporation of nanoparticles into a scaffold. The particles employed can be structurally classified as micelles, dendrimers, liposomes, solid-lipid nanoparticles, and polymeric nanoparticles [3, 4, 7–11]. The size range of particles used in controlled release varies between 20 and 1000 nm and impacts the release kinetics of the loaded bioactive molecule. Among these systems, polymeric nanoparticles have several advantages, including high stability, biodegradability, and flexibility in regulating the release rate of the GFs. Both the loaded particles and cells are incorporated into the biomaterial matrix that is used to fabricate the tissue scaffold.

## 8.5  Controlled Release via Micro/Nanoparticles

### 8.5.1  Techniques to Prepare Particles

Various techniques have been investigated to prepare micro/nanoparticles for use in controlled release systems. Emulsion solvent evaporation techniques including single and double emulsions are commonly used due to the ease of preparation and their applicability to a wide range of materials, while microfluidics is a growing field of interest for the preparation of particles due to the greater degree of control it offers. Both natural and synthetic materials have been used to prepare particles using these methods. A non-exhaustive list of such materials includes chitosan, alginate, PLA, PCL, and poly(ethylene glycol) (PEG). Organic solvents including dichloromethane (DCM), chloroform, and ethyl acetate are often used as the organic phase, with selection based on their ability to dissolve the encapsulating polymer. Surfactants (gelatin, albumin, ethyl cellulose, etc.) are added during emulsion techniques to stabilize the various layers of prepared particles.

Single emulsion techniques are commonly based on oil-water solvent evaporation, as shown in Fig. 8.8, and are highly efficient for encapsulation of water-insoluble molecules. The encapsulating polymer is dissolved in the organic phase, with mixing to ensure dispersion. The bioactive molecule to be incorporated is also added to this organic phase. The aqueous phase is then prepared by adding a surfactant. Emulsification is then carried out by adding the organic phase to the aqueous phase in a drop-wise fashion while the aqueous phase is vigorously agitated. This limits the interaction of organic droplets dispersed in the aqueous phase, allowing the surfactant to stabilize individual particles and limit the size of the particles formed. The solvent is then removed through evaporation, often using a rotary evaporator. The particles formed are then centrifuged and washed with distilled water to remove any remaining surfactant or free bioactive molecule. These particles are then lyophilized (freeze-dried) and collected.

The water-in-oil-in-water double emulsion technique, as shown in Fig. 8.9, can be used to prepare microparticles loaded with GFs, therapeutic agents, or other bio-molecules and overcomes issues with single emulsion techniques such as lack of control over release of highly soluble compounds. To prepare the particles using a double emulsion solvent evaporation technique, the biomolecules are first suspended

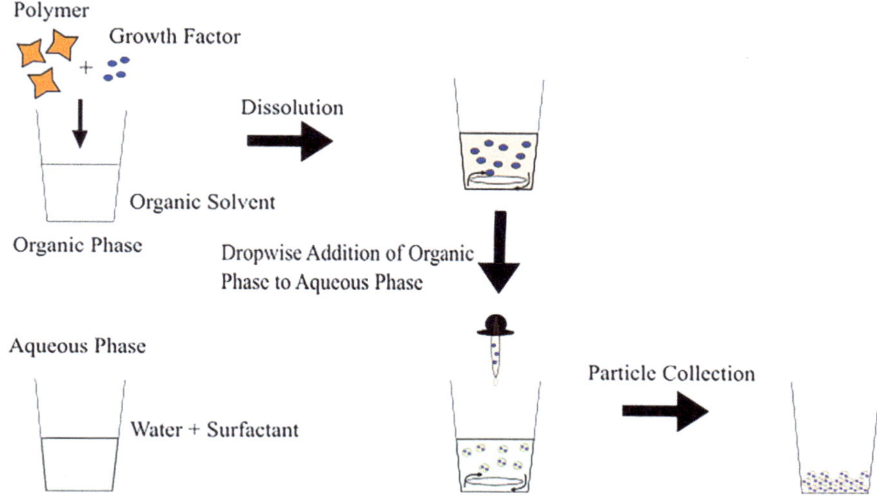



**Fig. 8.8**  Schematic diagram of oil-in-water emulsion

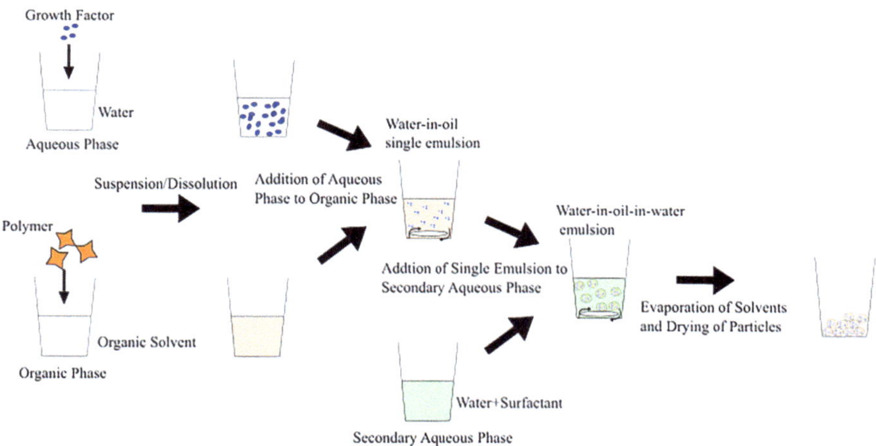



**Fig. 8.9**  Schematic diagram of water-in-oil-in-water double emulsion

in an aqueous solution. The microparticle material (polymer) is dissolved in an organic phase that is insoluble in the aqueous phase. These two solutions are vigorously mixed to form a water/oil emulsion. This primary emulsion is then mixed with a secondary aqueous solution that contains surfactants/emulsifiers to ensure the primary emulsion remains intact. The organic phase can then be removed via evaporation, leading to the formation of stable dual-layer microparticles [12]. The microparticles are then filtered out and washed to remove any remaining emulsifier using similar techniques to those discussed for single emulsion.

In some cases of double emulsion, a solvent extraction technique may be used instead of a solvent evaporation technique. This removes the solvent present in the external phase and requires the addition of the double emulsion to a third solution in which the polymer is insoluble but the organic and aqueous phases are miscible. This extracts the solvent present in the polymer particles into the aqueous medium, thus forming solid particles instead of particle shells. Other double emulsion techniques such as solid/oil/water (internal solid phase encapsulated in an aqueous phase through separation by an oil layer), solid/oil/oil (solid phase encapsulated in two non-miscible oil phases), and water/oil/oil (aqueous phase encapsulated in two non-miscible oil layers) emulsions have also been investigated for increased protein stability and further tailoring of release kinetics depending on the characteristics of the selected polymer and bioactive molecules.

Microfluidics has been gaining popularity in the last decade as the technology has advanced. Droplet microfluidics allows for the precise production of droplets between 1 and 1000 μm in diameter with a high degree of repeatability from a wide range of materials [13]. While controlling the size, dispersion, and shape of particles formed through the various emulsion techniques is difficult, microfluidics allows for tuning of particle composition and geometry. Droplets are generated using microfluidic devices (commonly glass capillary or lithographically fabricated poly (dimethylsiloxane) (PDMS) devices) from single or double emulsions. Drops are fabricated one at a time, with applied forces carefully balanced to control fluid behavior and droplet formation. Channel dimensions limit the size of the formed droplet, which is further controlled through hydrodynamic forces in the chip, allowing for diameters as small as hundreds of nanometers [13] (Fig. 8.10).

To form a water/oil emulsion, co-axial injection of the aqueous and organic phases in the presence of a hydrophobic surface leads to the encapsulation of the aqueous phase. For an oil/water emulsion, the internal surface of the microfluidic device is rendered hydrophilic, leading to the encapsulation of the organic phase by the aqueous phase. Droplets are formed at the capillary orifice and dropped onto a substrate for collection. By manipulating the injection geometry of the fluid phases, particles of different sizes can be formed. Double emulsions are formed in a similar way to single emulsions; once the single emulsion has been formed, a third immiscible phase is introduced, and the surface of the microfluidic device is treated to achieve the desired hydrophilicity, allowing the formation of a double-layered droplet. Higher-order emulsifications may also be formed through sequential emulsifications at junctions on the microfluidic device [13].

## 8.5.2  *Techniques to Characterize Controlled Release*

When developing a controlled release system, characterizing the morphology of the encapsulating materials/particles, encapsulation capacity, loading efficiency, release profile, and related bioactivity in the host system is important so as to determine if the system is providing the desired results.

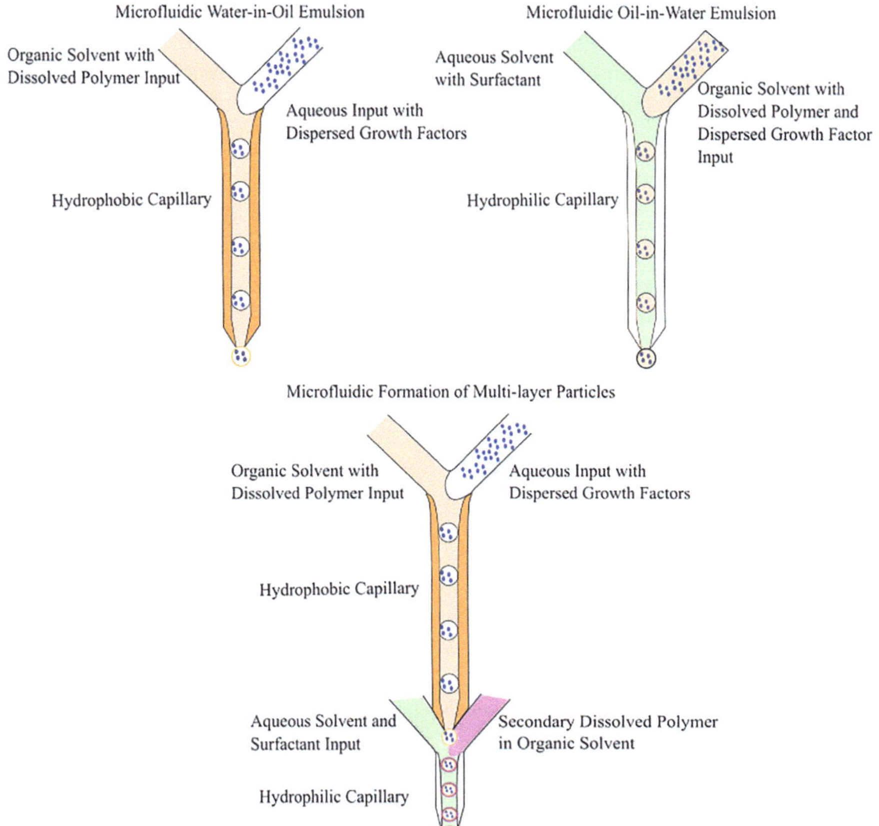


**Fig. 8.10**  Schematic diagram of microfluidic fabrication of microparticles

For particle encapsulation systems, morphological characterization of the encapsulating particles is used to determine the particle size, shape, and distribution as these factors will affect the release rate and profile. Particles that have a larger surface-area-to-volume ratio will allow for faster release of loaded components while degrading at a relatively higher rate than particles with lower surface-area-to-volume ratios. Distribution in terms of size and shape leads to inconsistency in release as each particle will hold varying amounts of the loaded bioactive molecules while degrading at different rates. Morphological characterization is accomplished through techniques such as scanning electron microscopy (SEM) and dynamic light scattering (DLS) [2, 5]. For SEM, freeze-dried particles can be mounted on conductive specimen holders and coated in gold to ensure conductivity. This allows for the visualization of the dimensions and morphology of the fabricated particles at high resolution. Using DLS, particles can be suspended in water, and the extent of light scattering used to quantify the particle mean diameter and size distribution for a larger sample size than can be characterized from individual SEM micrographs.

Loading/encapsulation capacity is a measure of the maximum concentration of bioactive molecules that can be incorporated into the matrix material/vessel. Maximum theoretical loading capacity ($\psi_{Theoretical}$) can be calculated using Eq. 8.1 [2]:

$$\psi_{Theoretical} = \frac{V_{IAP} \times C_{biomolecule}}{V_{OP} \times C_{vessel}}. \tag{8.1}$$

Actual loading capacity is experimentally determined by collecting the supernatant of the loaded vessel after undergoing a stimulus that causes the release of all loaded biomolecules, e.g., a thermal stimulus (incubation) or mechanical stimulus (vibration, agitation). The concentration of bioactive molecules released into the supernatant is recorded and divided by the concentration of the loading material (Eq. 8.2) [2]:

$$\psi = \frac{C_{biomolecule}}{C_{vessel}}. \tag{8.2}$$

To aid in measuring the concentration of bioactive molecules, fluorescent labeling may be used before these molecules are loaded to allow for characterization using techniques such as ultraviolet (UV) spectrophotometry, infrared imaging, or other means of quantifying fluorescence [2, 5]. Enzyme-linked immunosorbent assay (ELISA) kits may also be used to quantify the concentration of bioactive molecules released in the supernatant.

Loading efficiency can then be calculated using Eq. 8.3 [2]:

$$\eta = \frac{\psi}{\left( \frac{V_{IAP} \times C_{biomolecule}}{V_{OP} \times C_{vessel}} \right)}, \tag{8.3}$$

where $V_{IAP}$ is the volume of the inorganic aqueous phase and $V_{OP}$ is the volume of the organic phase in the loading system. Loading efficiency can also be calculated as a mass percentage, in which the mass of the incorporated bioactive molecules is divided by the total mass of the particle.

As previously discussed, the release profile is influenced by the mechanism of release as well as the physiochemical, architectural, and structural characteristics of the encapsulating vessel. The release profile obtained for a specific controlled release system can be characterized through techniques similar to the loading capacity; however, the bioactive molecule concentration present in the supernatant is quantified at various time points, which allows for percent release over time to be plotted in reference to the loading capacity.

Bioactivity can be inferred from the concentration of the bioactive molecule present in the supernatant; however, characterization of bioactivity commonly takes place in vivo or in assay systems meant to mimic physiological conditions.

Characterization of bioactivity often includes analysis of the physiological impact of the release, which is influenced by concentration as well as release kinetics, with the duration of sustained concentrations in an effective dose playing a major role in physiological impact. For example, bioactivity of angiogenic factors may be assessed using various microscopy techniques based on the length of new blood vessels produced [2]. In the case of immunostimulatory bioactive molecules, immunohistochemical techniques, such as ELISA, may be used to determine the degree of production of certain immune signaling molecules [5].

Mathematical modeling has also been employed to predict and design vessels with certain release profiles. These mathematical approaches commonly consist of a series of differential equations that account for different conditions and designs, such as vessel size and design, polydispersity index, matrix concentration, loaded concentration of the bioactive molecule, decomposition rate of the matrix and bioactive molecule, matrix porosity, and loading capacity, among other influential factors [3]. Verification of a mathematical model with experimental data allows for the more efficient design of future controlled release systems.

### 8.5.3 Rate Programming of Particles

Rate programming of particles aims to control the dose, sequence, and profile of the release of incorporated bioactive molecules so as to regulate cellular activities and the interactions between the cells and matrix material during tissue regeneration. This is achieved through fabrication of particulate delivery systems with well-understood physiochemical/ architectural modulated mechanisms of release [5, 10, 11, 14–16]. Due to the increased tailorability of synthetic materials in relation to physical, mechanical, and degradation properties, they are often preferred for application in pre-programmed release systems, with the incorporation of cells and bioactive molecules helping to increase their biocompatibility.

Chemical properties, including hydrophilicity, molecular weight, viscosity, degree and mechanism of cross-linking, crystallinity, and presence/number of functional groups, all play a role in the release rate and release profile from the matrix material. Further, degree of loading, porosity, and average pore size of the encapsulating particle are critical factors in bioactive molecule release from an encapsulating particle system. Structural and architectural properties including particle size, size distribution, number of shells, and material selection all influence the loading capacity and loading efficiency, which also play critical roles in determining release characteristics. These properties can be influenced by the selected fabrication method (single emulsion, double emulsion, etc.) and the ratio and concentrations of polymer and organic and aqueous phases. Complex mathematical models, often consisting of a series of differential equations relating release characteristics to physical and chemical properties, can be used to optimize the fabrication conditions for a pre-programmed particulate system, accounting for factors such as number and thickness of shells and material selection. These properties and selection are based

**Table 8.2** Comparison of release mechanisms and critical influencing factors

| Release mechanism | Influencing factors |
| --- | --- |
| Diffusion controlled | Polymer mesh size |
| | Growth factor molecule size |
| | Material/structure porosity |
| | Technique used for loading biomolecules |
| Homogeneous degradation | Swelling/hydration |
| | Polymer molecular weight |
| | Particle size |
| | Crystalline/amorphous matrix |
| Surface degradation | Matrix surface area |
| | Polymer molecular weight |
| | Technique used for loading biomolecules |

on understanding both the mode(s) of release (diffusion controlled, surface degradation, etc.) and loading mechanism (chemical immobilization, physical encapsulation, etc.) as these further influence the release profile, as previously discussed.

Core-shell systems in which the core contains the incorporated bioactive molecule have also been designed. The shell functions as a rate-limiting barrier, leading to delayed release. This delay is tailored through the selection of a shell material with desired chemical properties to achieve an appropriate degradation rate, as well as the fabrication of a shell with desired physical characteristics such as thickness to obtain the desired duration of delay. Double-shelled systems can incorporate multiple GFs for release at different time points, as degradation of the outer shell allows the release of bioactive molecules loaded in the middle material. Subsequent degradation of the middle material will reveal the core material, which will also begin to degrade and thus release the second loaded molecule.

A summary of the various release mechanisms and their corresponding release profiles is provided in Table 8.2.

**Case Study 8.1** Alginate microspheres loaded with immunoglobulin G (IgG) were prepared using a water-in-oil single emulsion technique and coated with chitosan polymer (Fig. 8.11) [5]. Sodium alginate (5% wt./vol. solution) was dispersed in paraffin oil with 5% Span 80 surfactant and stirred vigorously for 1 h at room temperature. Calcium chloride cross-linking solution was slowly added to the emulsion to promote gelation of the alginate microspheres. Isopropyl alcohol was then added to harden the microspheres before washing, centrifugation, and lyophilization. Both small-scale volumes (microliters) and larger-scale preparation volumes were tested for optimal microsphere preparation, along with different stirring speeds. The smaller-scale technique resulted in smaller particles with a more consistent morphology, with higher stirring speeds causing a reduction in mean particle diameter. Freeze-dried prepared particles were then coated with chitosan through cationic adsorption. Optimized microspheres were further incorporated into alginate scaffolds to achieve the prolonged release of protein for tissue engineering

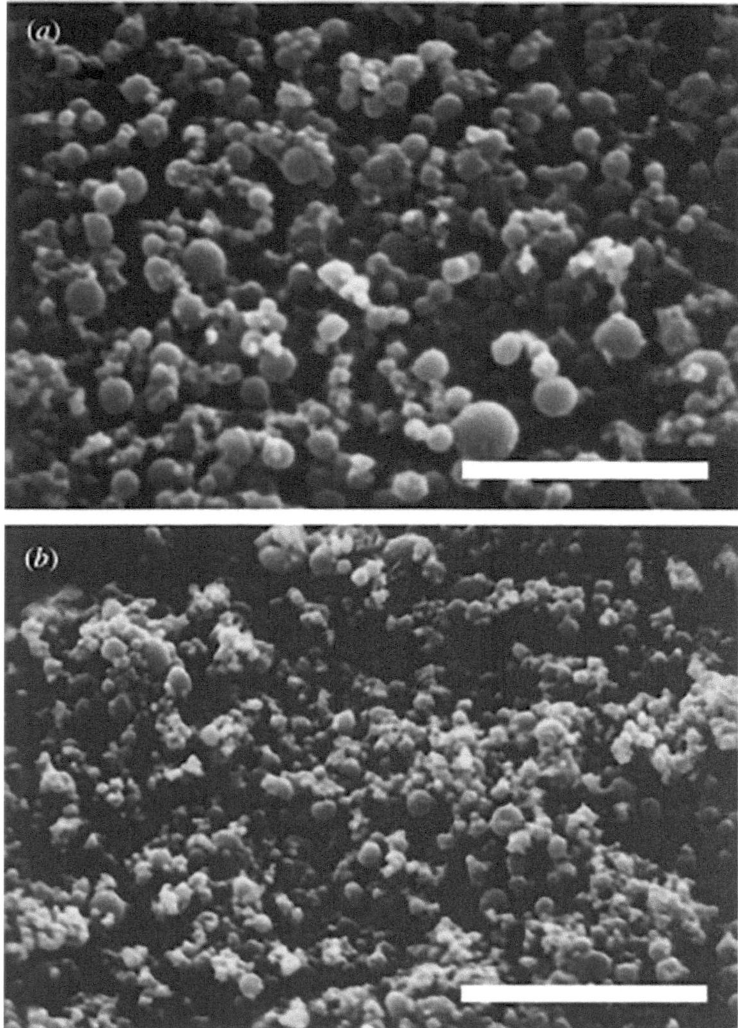

**Fig. 8.11** SEM micrographs of (**a**) alginate microspheres and (**b**) chitosan-coated microspheres prepared using the small-scale emulsion technique at a stirring speed of 1300 rpm. Scale bar = 10 μm. Used with permission from [5]

applications. The secondary coating with chitosan reduced the initial burst release and subsequent protein release compared to uncoated alginate microparticles. The release of the protein was observed to be controlled by the properties of both the microsphere and scaffold, including the degradation rate, swelling degree, and diffusion of the protein into the hydrogel scaffold.

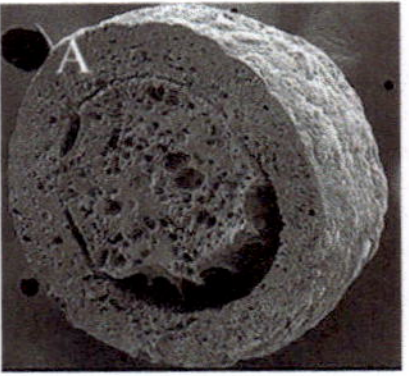

**Fig. 8.12** Release of growth factors from bi-layer polymeric nanoparticles to promote angiogenesis using a rat aortic ring assay: (**a**) SEM image of a bilayer particle composed of a poly(L-lactide) (PLLA) shell and a poly(lactide-co-glycolide) (PLGA) core; angiogenic response of aortic rings at day 8 to (**b**) VEGF release alone, (**c**) co-delivery of VEGF/bFGF, and (**d**) co-delivery of VEGF/bFGF followed by the release of PDGF (scale = 1000 μm) [2]

**Case Study 8.2** Polymeric core-shell (or bilayer) nanoparticles for synchronized and sequential release of several GFs (e.g., VEGF, bFGF, PDGF) were synthesized using a water-in-oil-in-water double emulsion method and incorporated in fibrin hydrogel to stimulate angiogenesis for cardiac tissue engineering [2]. The biolayer nanoparticles reduced the initial burst release to less than 5% and delayed the release of GFs over 20 d, during which the integrity of the entrapped GFs was maintained (see Fig. 8.12). The bi-layer nanoparticles allowed for sequential GF release, i.e., co-delivery of VEGF/bFGF followed by the release of PDGF; this sequential GF release was more effective for promoting angiogenesis than the delivery of VEGF alone or in combination with bFGF (Fig. 8.12), as evaluated using a rat aortic ring assay.

**Case Study 8.3** Hepatocyte growth factor (HGF)-loaded chitosan-coated alginate nanoparticles ~1000 nm in diameter were synthesized using a water-in-oil-in-water emulsion technique before being incorporated into an alginate/collagen hydrogel [17]. 3D tissue scaffolds containing primary human pulmonary fibroblasts were then bioprinted from this nanoparticle-containing hydrogel, and constructs were seeded with primary human bronchial epithelial cells. Release of HGF was sustained over a 14-day period, with the chitosan coating lessening the initial burst release. The effects of culture conditions on release rate were investigated by comparing standard incubation conditions with breath-mimicking incubation conditions in a bioreactor. It was determined that the biomechanical forces imposed by the bioreactor did not have a significant impact on release rate. Further evaluation was carried out on cell metabolism between scaffolds with and without nanoparticles. While cell metabolism remained consistent between the control and nanoparticle-loaded scaffolds, confocal imaging suggests that epithelial proliferation and differentiation were upregulated, with a more consistent epithelial barrier layer formed in nanoparticle-containing scaffolds (Fig. 8.13).

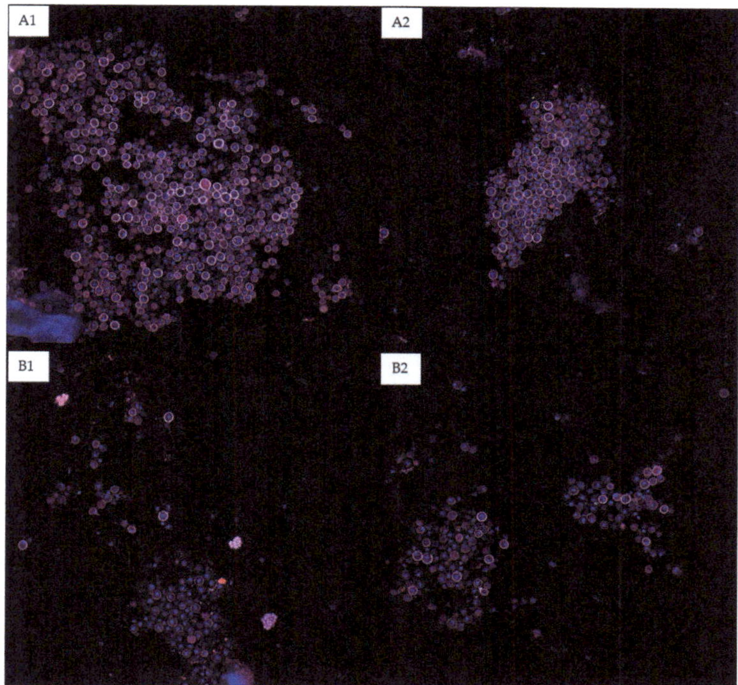

**Fig. 8.13** Confocal images of the epithelial surface layer of the bioprinted constructs at the Day 14 timepoint. (**A1, A2**) Scaffold containing 4 µg/mL HGF-loaded nanoparticles, (**B1, B2**) control [17]

## 8.5.4   *Actively Controlled Release*

In addition to pre-programmed controlled release based on the chemical and physical properties of the fabricated system, the use of external and internal stimuli to actively control release is also being investigated. Internal stimuli such as pH and chemical/ biological concentration and/or externally controlled stimuli such as temperature and light exposure can trigger significant conformational changes in some polymer materials, making them likely candidates for actively controlled release systems.

Cellular activities can produce local acidic or basic conditions through enzymatic reactions. In the case of pH-sensitive systems, once a microenvironmental pH threshold is reached, the delivery polymer is designed to undergo a conformational change that allows for the release of incorporated bioactive molecules. When developing pH-sensitive systems, the matrix material is selected to be stable at physiological pH while the cellular activities localized on the implanted scaffold change the pH in the microenvironment to trigger bioactive molecule release. Polymers such as poly(acrylic acid) (PAA), poly(methacrylic acid)(PMAA), poly(2-ethyl acrylic acid), and poly(2-propyl acrylic acid) are pH-responsive materials that are currently being investigated for use in actively controlled pH-dependent release systems [3]. Stimuli including the concentration of certain biologically relevant materials, such as glucose, that change related to the state of the host/host cells can also be used to trigger active release systems through interactions with functional groups on the polymer backbone [15].

Thermoresponsive polymers can be used in thermally controlled active release systems, as they undergo phase changes at either an upper or lower critical solution temperature that cause the polymer to become miscible with its microenvironment or to precipitate. Techniques such as magnetic resonance, radio frequency waves, microwaves, and ultrasound can be used to increase the local temperature of thermosensitive polymers, inducing a conformation or phase change through an external stimulus. Heating of the polymers can increase the miscibility or pore size of the material, allowing the release of incorporated bioactive molecules. Multiple thermosensitive materials with different lower critical solution temperatures can be used in these systems to achieve multiple release profiles, as different energy inputs will be required to trigger release from individual materials. Precipitation of polymers may also trigger release, as this increases the internal pressure inside particles and causes them to osmotically pump out incorporated biomolecules. Polymers such as PLGA-PEG-PLGA, poly(N-vinylcaprolactam) (PNVCL), and phosphonomethyl-pentanedioic acid (PMPA), among others, have been investigated for use in thermally responsive controlled release systems [3].

UV or visible light can also be used as an external stimulus to activate release in some materials. UV-sensitive materials can undergo a phase change and corresponding volume change when exposed to an external stimulus, causing the release of entrapped molecules. Light-absorbing chemophores that are responsive to visible light may also be incorporated into matrix or particle systems along with bioactive molecules. In these cases, exposure to visible light of specific wavelengths and intensities can generate heat within the chemophore, triggering the release of bioactive molecules from thermosensitive materials by increasing swelling of the material. Unfortunately, UV and visible light exhibit low penetrative ability. To activate light sensitive systems, near-infrared radiation (NIR) with a wavelength between 650 and 900 nm can be used to increase penetration. Gold nanosystems such as gold nanorods, shells, and cages can be used as components in NIR active release systems as they are efficient absorbers of NIR.

## 8.6  Design of a Controlled Release System

Controlled release systems must be tailored to specific tissue engineering applications. Understanding the physiologically relevant biomolecules, mode of release of these bioactive molecules from a selected biomaterial, and different possible loading mechanisms helps facilitate the design of controlled release systems. The first step involves selection of the bioactive molecules to be loaded into the matrix material or encapsulating particles. This selection is based on an understanding of the physiological system, including what concentration of the molecules is required to achieve the desired degree of bioactivity. Understanding the effective therapeutic window in terms of both dose and required duration informs the selection and design of the encapsulating particles or matrix materials. For example, if a quick release of high concentration is required, a diffusion-controlled system loaded through physical encapsulation may be preferred, whereas an application requiring extended release may be best achieved by a system loaded through chemical immobilization and released through surface degradation.

The second step is to select or determine the methods to load biomolecules into the material or encapsulate particles. These methods include chemical immobilization, physical encapsulation, or loading in micro/nanoparticle systems, as discussed in Sect. 8.4. Understanding the loading capacity (maximum amount of bioactive molecule that can be incorporated into a particle or matrix) and loading efficiency (percent of the loading capacity that can be incorporated into the particles or matrix based on the loading techniques and/or mechanisms) allows for the selection of dosage, as a system with a high loading capacity will be able to deliver a larger concentration of bioactive molecule in a smaller amount of particles/volume of matrix material.

Various release profiles have been achieved through different systems, with the most common being burst release, zero-order kinetics, or multi-release profiles. Burst release profiles are used when quick onset is required and may require the design of system with maximized surface area for enhanced diffusion and degradation. Zero-order release profiles demonstrate a relatively constant release rate and are designed to sustain the bioavailability of the molecule within the therapeutic window over an extended duration. These release profiles reduce the incidence of adverse effects and are generally well tolerated by the host system. Multi-release profiles are often used when incorporating more than one biomolecule is desirable, with the biomolecules released at different time points, or when designing an immunologically active scaffold, with the second release functioning as a booster for immune memory. Control over timing of release is an important consideration when designing a controlled release system for optimal bioactivity, as one GF may stimulate migration and proliferation and a secondary GF may stimulate differentiation and cellular maturation, as is the case for angiogenesis.

## 8.7  Summary

In tissue engineering, various biomolecules such as GFs, adhesion factors, and immunogenic materials can be used to promote and control cellular proliferation, migration, and differentiation. By incorporating them into engineered tissue scaffolds, the release of these biomolecules can help to stimulate intracellular signaling cascades that promote the regeneration of tissues, thus helping the tissue to recover its functionality.

The mode of release of such biomolecules depends on the interactions between the loaded biomolecules, matrix material, and external environment. Typical mechanisms of release are diffusion-controlled, homogeneous degradation, surface erosion, or immobilization modes with varying release profiles. The methods to load biomolecules include chemical immobilization, physical encapsulation, and use of particle systems. These loading methods can further influence the release kinetics and loading capacity/efficiency of the controlled release system. Mode of release and mechanism of loading are important considerations to ensure the desired release profile is achieved, whether that be prolonged release within the desired therapeutic window for maximum beneficial effect, a fast one-step profile, or a multi-stage release profile.

Particle delivery systems can be achieved using different methods, commonly including single emulsion, double emulsion, or microfluidic techniques. While single and double emulsion techniques are simple and allow for tailoring of release profiles through layer selection, microfluidic technologies provide more homogeneous particles and the ability to form consistent multi-layered particles. These particle systems can deliver multiple different incorporated molecules in subsequent order by varying the layers that contain particles. To program particle release at pre-determined times, factors including the size of particles, layer thickness, and layer composition must be taken into account. This is often done through the use of verified mathematical models of release system designs. Active release systems may also be used to control biomolecule release by selecting materials that react to either internal or external stimuli, such as thermosensitive, light sensitive, pH sensitive, and/or concentration sensitive polymers.

When designing a controlled release system for a specific tissue engineering application, necessary considerations include the selection of biomolecules and biomaterials as well as the structural design of the encapsulating particles and/or matrix material. Increased understanding of the role of biologically active molecules, such as GFs, and their incorporation into tissue scaffolds can increase successful applications of controlled release systems in tissue regeneration and other applications such as immunological mediation.

**Problems**

1. Explain the concept of controlled release of biomolecules in tissue engineering and its purpose.
2. Identify a study from the recent literature to briefly explain why the controlled release of biomolecules is needed in vitro and/or in vivo.
3. Name two biomolecules and briefly describe their use in tissue engineering.
4. Briefly explain the modes of controlled release.
5. Briefly explain the methods to load biomolecules into scaffolds.
6. Briefly explain one technique to prepare particles for GF controlled release, with the help of an illustration.
7. Briefly explain one technique to characterize particles for GF controlled release, with the help of an illustration.

# References


1. J. Rice, M. Martino, E. Scott, et al., Controlled release strategies in tissue engineering, in *Tissue engineering*, (Elsevier, 2015), pp. 347–392
2. M. Izadifar, M.E. Kelly, X. Chen, Regulation of sequential release of growth factors using bilayer polymeric nanoparticles for cardiac tissue engineering. Nanomedicine **24**, 3237–3259 (2016). https://doi.org/10.2217/nnm-2016-0220
3. M. Izadifar, A. Haddadi, X. Chen, et al., Rate-programming of nano-particulate delivery systems for smart bioactive scaffolds in tissue engineering. Nanotechnology **26**, 012001 (2015). https://doi.org/10.1088/0957-4484/26/1/012001
4. R. Fang, S. Qiao, Y. Liu, et al., Sustained co-delivery of BIO and IGF-1 by a novel hybrid hydrogel system to stimulate endogenous cardiac repair in myocardial infarcted rat hearts. Int. J. Nanomedicine **10**, 4691–4703 (2015). https://doi.org/10.2147/IJN.S81451
5. P. Zhai, X.B. Chen, D.J. Schreyer, et al., Preparation and characterization of alginate microspheres for sustained protein delivery within tissue scaffolds. Biofabrication **5**(1), 015009 (2013). https://doi.org/10.1088/1758-5082/5/1/015009
6. A. Zimmerling, X. Chen, Bioprinting for combatting infectious diseases. Bioprinting **20**, e00104 (2020). https://doi.org/10.1016/j.bprint.2020.e00104
7. P. Zhai, X.B. Chen, D.J. Schreyer, An in vitro study of peptide-loaded alginate nanospheres for antagonizing the inhibitory effect of Nogo-A protein on axonal growth. Biomed. Mater. **10**(4) (2015). https://doi.org/10.1088/1748-6041/10/4/045016
8. P. Zhai, X.B. Chen, D.J. Schreyer, et al., PLGA/alginate composite microspheres for hydrophilic protein delivery. Mater. Sci. Eng. C Mater. Biol. Appl. **56**, 251–259 (2015). https://doi.org/10.1016/j.msec.2015.06.015
9. F. Mohabatpour, M. Al-Dulaymi, L. Lobanova, et al., Gemini surfactant-based nanoparticles T-box1 gene delivery as novel approach to promote epithelial stem cells differentiation and dental enamel formation. Biomater. Adv. **137**, 212844 (2022). https://doi.org/10.1016/j.bioadv.2022.212844
10. M. Izadifar, M.E. Kelly, X. Chen, Computational nanomedicine for mechanistic elucidation of bilayer nanoparticle-mediated release for tissue engineering. Nanomedicine **12**(5), 423–442 (2017). https://doi.org/10.2217/nnm-2016-0404
11. M. Izadifar, M.E. Kelly, A. Haddadi, et al., Optimization of nanoparticles for cardiovascular tissue engineering. Nanotechnology **26**(23), 235301 (2015). https://doi.org/10.1088/0957-4484/26/23/235301
12. T. Giri, C. Choudhary, et al., Prospects of pharmaceuticals and biopharmaceuticals loaded microparticles prepared by double emulsion technique for controlled delivery. Saudi Pharm. J. **21**, 125–141 (2013). https://doi.org/10.1016/j.jsps.2012.05.009
13. W. Li, L. Zhang, X. Ge, et al., Microfluidic fabrication of microparticles for biomedical applications. Chem. Soc. Rev. **47**(15), 5646–5683 (2018). https://doi.org/10.1039/c7cs00263g
14. F. Yasmin, X. Chen, B.F. Eames, Effect of process parameters on the initial burst release of protein-loaded alginate nanospheres. J. Funct. Biomater. **10**(3), 42 (2019). https://doi.org/10.3390/jfb10030042
15. R. Cheng, F. Meng, C. Deng, et al., Dual and multi-stimuli responsive polymeric nanoparticles for programmed site-specific drug delivery. Biomaterials **34**, 3647–3657 (2013). https://doi.org/10.1016/j.biomaterials.2013.01.084
16. B. Zhang, Y. Zhao, K. Guo, et al., Macromolecular nanoparticles to attenuate both reactive oxygen species and inflammatory damage for treating Alzheimer's disease. Bioeng. Trans. Med. **8**(3), e10459 (2022). https://doi.org/10.1002/btm2.10459
17. A. Zimmerling, C. Sunil, Y. Zhou, et al., Development of a nanoparticle system for controlled release in bioprinted respiratory scaffolds. J. Funct. Biomater. **15**(1), 20 (2024). https://doi.org/10.3390/jfb15010020

# Index


**A**

Actively controlled release systems, 231–232

Advanced bioprinting, 11–12, 141, 179–183

Agarose, 11, 40, 41, 45–47, 53, 102, 103, 110, 114, 162, 206

Alginate and gelatin composite hydrogel, 51

Alginate hydrogel, 91, 181

Alginate solutions, 45–47, 81, 91, 132–135, 155, 182, 203

Angiogenesis, 12, 192–194, 196, 209, 215, 216, 230, 233

Anisotropy, 82–84, 102

Architectural properties, scaffolds, 2, 7, 13, 17–21, 25

Architecture, 3, 4, 7, 17, 18, 23, 25–26, 31–33, 51, 98

**B**

Basics of solution preparation, 109–114

Bending tests, 67–69, 105
  biocompatibility, 42–43

Biodegradation, 42–44, 50, 52

Bioink, 9–11, 13, 29, 33, 37–53, 114, 120–122, 132, 135, 141, 143–155, 158–162, 164–168, 171, 172, 175, 176, 179–186, 191, 195–196, 198, 206, 207, 210

Bioink flow, 144–148, 151, 152, 161, 171, 183, 184, 186

Biological properties, 2–4, 11, 13, 23–26, 31–33, 37, 42–44, 47, 98

Biomaterials, 1–5, 7–13, 17, 21–24, 26–33, 37–53, 72, 79, 97, 102, 104, 109–137, 141, 179, 181, 182, 186, 195, 204, 216, 217, 219, 220, 222, 232, 234

Biomaterial solution, 8–10, 12, 13, 38–40, 47, 109–137, 141, 186

Biomolecules, 2, 109, 141, 207, 213–234

Bioprinted structures, 27

Bioprinting, 1–13, 17, 18, 25–27, 29–34, 37–53, 110, 114, 115, 117–120, 132, 133, 136, 137, 141–187, 191–210

Bioprinting parameters, 164–166, 184

Bioprinting process, 10, 13, 37–42, 45, 109, 117, 119, 132, 133, 136, 137, 141, 144–146, 152, 154, 162, 164–166, 169–171, 174–178, 185, 186

Bioprinting vascular networks, 196–207, 209

Blood vessel, 46, 48, 191–197, 199, 206, 208–210, 215, 227

**C**

Capillary rheometer, 124–127, 136

Carreau fluid model, 119, 121, 136

Cartesian coordinate system, 84

Casson model, the, 119, 121, 136

Cell bioprinting, 114, 133

Cell damage, 40–42, 143, 170–179, 184–186, 208

Cell viability, 10, 11, 13, 26, 42, 45, 47, 50, 52, 53, 114, 133, 141, 157, 170, 174–179, 181, 184–186, 202, 204

Chemical immobilization, 219–220, 228, 233, 234

Chitosan, 10, 43, 46, 98, 100, 110, 114, 137, 162, 170, 195, 208, 222, 228–230

Coaxial bioprinting, 182, 185, 196, 198
Collagens, 10, 11, 29, 31, 40, 41, 45, 47–49,
    51–53, 84, 98–100, 102, 110, 114, 162,
    164, 195, 208, 230
Collagen scaffolds, 48
Complex modulus, 75, 103
Composite material, 51, 98–101
Composite scaffolds, 99–100
Computed tomography (CT), 4, 24, 25, 33, 169
Cone-and-plate rheometer, 124, 127–129, 135,
    136, 185
Controlled release, 213–234
Controlled-release modes, 213, 215–219
Controlled release systems, 214, 219, 222, 224,
    226, 227, 232–234
Cortical bone specimen, 65, 85, 86
Creep, 57, 72–75, 77, 78, 80, 81, 102, 103
Creep and relaxation testing, 73–75
Crosslinking, 10, 12, 37–41, 43, 45–49, 52, 53,
    81, 91, 131, 132, 161–164, 179, 180,
    182, 184, 186, 200, 202
Cyclic loading test, 73

**D**
Decellularized matrix (dECM) materials, 49,
    110, 114, 209
Design parameters, 17, 30, 33
Digital imaging and communications in
    medicine (DICOM), 26
Direct bioprinting, 196–200
Direct bioprinting of a vascular network,
    196–200
Directional dependency, 83
Dulbecco's modified Eagle's medium
    (DMEM), 132, 200, 207
Dynamic testing, 72–82, 91, 103
    ECM secretion, 204

**E**
Elastic modulus, 21, 51, 62, 63, 65–69, 76, 86,
    89, 90, 92, 94, 97, 98, 100, 102–105
Electrospinning, 5–7, 13, 100, 102, 183, 208,
    209
Electrospinning fabrication techniques, 6, 13,
    102, 183
Ellis fluid model, 119–121, 136, 137
Embedded bioprinting, 11, 161, 179–180, 185
Extensional stress, 10, 170–172, 174–178, 185
Extensometer, 62, 85
Extrudability, 38, 52, 157–158, 183, 186

Extrusion-based bioprinting techniques, 9, 11,
    18, 29, 179–183, 185, 187
Extrusion bioprinting, 1–13, 37, 42, 45, 47, 49,
    115, 117, 141, 143, 148, 183, 209
Extrusion printing, 8, 9, 38
    extrusion-based (EB) system, 141, 142, 172,
    185, 186, 206

**F**
Fibrin, 11, 40, 41, 45, 48, 49, 53, 102, 103, 110,
    114, 160, 195, 208, 220, 230
Fibrin-based scaffolds, 49
Fibrinogen, 48, 49, 215
Fillers, 11, 47, 51, 101, 104
First layer, printed bioink, 39
Flexural elastic modulus, 67, 68, 105
Flow behavior, 10, 39, 40, 50, 52, 109–137,
    144–146, 148, 153–155, 157, 158, 164,
    175, 179, 183–186
Flow behavior models, 119–124
Flow rate, 125, 126, 133, 144–160, 165, 176,
    177, 182–186, 198, 203, 207
Freeze drying, 5, 6, 13, 100, 102
Frequency-dependent test, 72, 75–76, 103

**G**
Gas foaming, 5, 6, 13
Gelatin, 11, 40, 41, 45, 48, 51–53, 98–100, 110,
    114, 117, 135, 162, 164, 180, 200, 208,
    222
Gelatin methacrylate composite (GelMA)
    hydrogels, 48
Generalized power-law model, 119–123, 136
Growth factors (GFs), 1, 12, 49, 98, 191, 193,
    194, 196, 199, 207, 209, 210, 213, 215,
    216, 220, 221, 228, 230

**H**
Hard tissue, 48, 49, 84
Herschel-Bulkley fluid model, 123
Human umbilical vein endothelial cells
    (HUVEC), 177
Hyaluronic acid (HA), 11, 45, 47, 48, 99, 100,
    110, 114, 180, 195
Hybrid bioprinting, 11, 179, 183, 185
Hybrid structure, 30, 33, 50, 102–104, 181
Hydrogel-based bioinks, 198
Hydrogel-based luminal tube, 204
Hydrogel-forming biomaterials, 45, 53

Hydrogel photopolymerization, 164
Hydrogel polymers, 12, 114
Hydrogels, 10, 12, 29, 30, 33, 37, 39–53, 72,
91, 102, 114, 126, 136, 161–164, 168,
169, 172, 181, 183, 184, 186, 191, 196,
198–200, 202–204, 206, 209, 210, 230

**I**
Imaging techniques, 24, 168–170, 184
Indirect bioprinting, 200–204
Infiltration of hydrogel, 102
Influence of cell density, 15
Influence of scaffold materials, 95–97
Influence of temperature, 133–134, 177
Ink-jet printing, 8

**K**
Kelvin-Voigt model, 76, 79–82, 103

**L**
Laser-assisted printing, 7, 8, 12, 13
Laser-induced forward transfer (LIFT), 8, 207
Layer-by-layer hydrogel printing, 8, 115, 144,
181
Linear viscoelastic behavior, 76–82, 130
Living cells, 3, 4, 7, 9, 10, 13, 17, 23, 33, 37, 42,
45–47, 52, 109, 114–115, 136, 141, 143,
170, 178, 181, 183, 184
Loading, 45, 57–62, 65, 68, 70, 72, 73, 75, 82,
83, 85, 89–94, 96, 103–106, 129, 204,
219–222, 226–228, 232–234
Loss angle, 75, 76, 106, 130–132
Loss modulus, 75, 76, 106, 131, 133

**M**
Magnetic resonance imaging (MRI)
matrix-assisted pulsed laser evaporation
direct writing (MAPLE DW), 207
Mass percent, 111
Mass/volume percent, 111
Material concentration, 132–133, 136
Maxwell model, 77, 78, 81, 82, 103
Mechanical force, 8, 10, 12, 21, 26, 72, 141,
143, 170–176, 185, 186, 230
Mechanical properties, 2, 11, 13, 21–23, 29–33,
37, 40, 43–49, 51–53, 57–106, 114, 136,
163, 164, 195, 199

Mechanical property measurements of native
tissues and scaffolds, 82–93, 104
Mechanical testing, 57–72, 83, 91, 103, 104,
106
Medical imaging, 4, 24–26, 33
Melt molding, 5, 6, 13
Mercury porosimetry, 168, 184
Micro-and nano-particles, 196, 213, 221–233
Micro-computed tomography (μCT), 168, 169,
184
Microfibrillated bacterial cellulose (MFC), 99
Microparticles, 135, 222, 223, 225, 229
Modeling, 1–4, 11, 13, 17, 23, 24, 26, 28, 43,
88, 175, 176, 227
Models, 2, 13, 24–26, 76–82, 88, 96, 103, 106,
115, 119–125, 127, 136, 137, 144, 155,
164, 169, 172, 175–177, 181, 184–186,
216, 227, 234
Mono-structure, 29, 30, 33
Multi-head, 11, 179, 181, 182, 185
Multiple-dispenser bioprinting, 182

**N**
Nanoparticles, 51, 101, 200, 221, 222, 230, 231
Native tissues and scaffolds, 57–106
Natural hydrogels, 45–49, 53, 114
Natural polymer, 29, 45, 46, 99, 195, 208
Needle geometry, 144, 154–157, 183, 186
Needle movement, 158–162, 164, 184, 186
Newtonian flow behavior, 115–116, 136
Newtonian fluid, 115, 116, 119, 125, 172
Non-Newtonian flow behavior, 116–117, 119,
125, 136
Non-soluble, 110, 135, 136

**O**
Operating temperature and humidity, 83, 104
Oscillatory rheometer, 124, 129–132, 136
Oscillatory shear technique, 129

**P**
Parallel plate rheometer, 124, 128–129, 136
Pearson's coefficient, 94, 96
Phase separation, 5, 6, 13, 100
Phosphate buffered saline (PBS), 83, 98, 206
Photo-cross linkable gelatin hydrogels, 48
Photo-curable polymers, 41
Physical encapsulation, 219–221, 228, 233, 234

Piston-driven bioprinting, 153–154
Pluripotent stem cells (PSCs), 28
Pluronic®
 poly(ethylene glycol) (PEG), 11, 50, 110,
  114, 200, 222
 poly(ethylene oxide) (PEO), 50, 110, 114,
  208
Pneumatic-driven bioprinting, 143–151, 157,
  164, 183
Polycaprolactone (PCL), 11, 29, 48–50, 88–90,
  94, 96–99, 102, 103, 110, 115, 181, 183,
  195, 208, 218, 222
Poly(ethylene)-based polymers, 50
Poly(lactic-co-glycolic acid) (PLGA), 48, 100,
  208, 218, 230
Pore geometry, 94, 168
Porogen leaching, 5, 6, 13
Porosity, 17–20, 30–34, 94–96, 99, 100,
  166–169, 184, 220, 227, 228
Power-law fluid model, 119–123, 125, 126,
  136, 137, 149, 155, 172, 185
Preparation of scaffold solutions, 113, 136
Printability, 10, 12, 13, 18, 37–41, 44, 48–50,
  52, 53, 132, 198, 200, 202
Printed hydrogel, 39, 43, 51, 181
Printed scaffolds, 9, 18, 22, 41, 44, 72, 85, 87,
  104, 109, 141, 144, 161, 164, 166–170,
  184, 213–234
Process-induced forces, 10, 12, 40, 170, 175

**R**
Rate programming, 227–231
Rheometers, 75, 124–132, 135, 136, 148, 177,
  185
Rhombicuboctahedron, 25

**S**
Sample mounting, 83, 104
Scaffold architecture, 23, 32, 33
Scaffold biological properties, 3, 4, 13, 23,
  31–33, 98
Scaffold design, 3, 4, 10, 13, 17–34, 88, 95, 97,
  114, 119, 158, 164, 165, 184, 186
Scaffold design and fabrication
 for compression testing, 88
Scaffold development, 4
Scaffold fabrication, 1, 4–8, 13, 21, 42, 47, 49,
  50, 97, 110, 133, 141, 143, 148, 160,
  183, 185
Scaffold materials, 2, 3, 18, 51, 95–97, 101,
  110

Scaffold mechanical properties, 2, 21–23, 32,
  57–106, 114, 136, 164, 196
Scaffold pore, 166, 169, 220
Scaffold requirements, 2, 3, 13, 17–23, 32, 33
Scaffolds, 1–13, 17–34, 41–52, 57–106,
  141–187, 213–234
Scaffold solution flow behavior, 114, 136
Scaffold structures, 6, 12, 17, 22, 23, 29–32, 44,
  94–95, 102, 109, 144, 166, 167, 186
Schwann cells, 3, 135, 155, 160, 174, 177
Screw-driven bioprinting, 143, 151–154, 183
Self-assembled vasculature, 204–207
Shear stress, 10, 39, 40, 42, 59–61, 70–72, 105,
  115–128, 130, 132, 133, 136, 137,
  146–149, 151, 170–179, 184, 185, 187,
  207, 208
Soluble material, 136
Solution preparation, 45, 46, 49, 109–114
Stereolithography, 26, 207
Storage modulus, 75, 76, 106, 131, 133
Strands, 8, 10, 18–20, 22, 29, 31–34, 39, 84, 88,
  94, 96, 97, 102, 141, 143, 144, 157–167,
  177, 182–186, 191, 196, 198–204, 206,
  209, 210
Stress-free (SF) speed, 159, 160, 184
Stress-relaxation test, 73–75, 78–81
Stress-strain behavior, 86, 89, 91
Stress-strain curve, 43, 44, 62–67, 73, 83, 86,
  88, 89, 91, 93, 96, 97, 99, 103
Stress-strain test, 75
Stromal cell-derived factor-1 (SDF-1), 194
Structural-theory models, 122–124, 136
Synthetic and composite hydrogels, 12, 42, 45,
  50–53, 114
Synthetic polymers, 11, 29, 30, 33, 37, 42, 48,
  51, 98, 102, 195, 208

**T**
Techniques to characterize flow behavior,
  124–132
Tensile and compressive tests, 62–67, 85, 86,
  102
Tensile test apparatus, 62
Tensile testing, 62, 65, 67, 83, 85–87, 102
 specimen preparation for, 85
TE scaffolds, 2–4, 6, 13
Thermal cross-linkable polymers, 41
3D printing, 5, 7, 8, 13, 26, 38
Thrombin, 48, 49
Time-dependent fluid flow behavior, 118, 136
Time-dependent generalized power-law model,
  121, 136

Time-dependent mechanical properties, 97–98
Time-independent fluid flow behavior, 117, 119, 136, 152
Tissue engineering (TE), 1–7, 11–13, 17, 23, 26–28, 30–32, 34, 41, 42, 47–52, 57, 79, 88, 91, 95, 97, 99, 101, 102, 104, 141, 160, 182, 208, 213–216, 228, 230, 232–234
Tissue modeling, 1–4, 13, 17, 23, 26
Tissue regeneration, 1, 3, 11–13, 21, 23, 29, 32, 33, 43, 44, 94, 97, 98, 103, 104, 179, 185, 191, 208, 213–215, 227, 234
Tissue scaffolds, 11, 13, 17–34, 47, 57–106, 109, 141–187, 191, 210, 213, 216, 222, 230, 233, 234
Torsion test, 69–72
Traditional techniques, 5–6, 8, 13
Tropocollagen molecules, 84
2D images, 24, 168
Typical tensile specimen, 62

**U**
Ultimate strength, 63, 65, 67, 86, 87, 104
Uniaxial compression test, 89
Universal testing machine (UTM), 88

**V**
Vascular endothelial growth factor (VEGF), 193, 194, 196, 199, 209, 215, 216, 230
Vascularization approaches, 207–209
Vascularization of scaffolds, 31, 47

Vascularized scaffolds, 191, 195, 197
Vascular networks, 12, 31, 191–210
Vasculature
    printed sacrificial networks, 201, 202, 210
Vasculogenesis, 192, 194–196, 209, 210
Vessel formation mechanisms, 191–195
Viscoelastic behavior, 76–82, 91, 103, 119, 129, 130, 133, 136
Viscoelastic fluid flow behavior, 119
Viscoelasticity, 72, 82, 102, 103
Viscoelastic properties, 72–82, 103, 106, 130, 135
Viscoplastic fluid flow, 117
Volume percent, 111

**W**
Water soluble and non-soluble, 40, 43, 45–47, 110, 114, 136, 192, 215, 216

**X**
X-ray absorption, 24

**Y**
Yield strength, 21, 63, 65, 67, 86, 96, 98, 99, 104

**Z**
Zonal structure, 17, 29–33, 183